THE BOURN DYNASTY

The Empire Mine's Golden Era

1869 – 1929

CHARLES C. STEINFELD

EMPIRE MINE PARK ASSOCIATION
1996

Acknowledgments

Photographs in this book were provided by the Empire Mine State Historic Park and Filoli. Produced by the Empire Mine Park Association's Publications Committee, Don Alexander, chairman, Jeanne Alexander, Roger Lescohier, and Jone Rothenberg.

ISBN: 978-1-57733-295-4
LCCN: 2015904993
Printed in the United States of America by
Blue Dolphin Publishing, P. O. Box 8, Nevada City, CA 95959
www.bluedolphinpublishing.com

About the Author

The author of this book, Charles C. Steinfeld, worked tirelessly for several years in researching and writing the story of the Bourn family, whose wealth and influence made the Empire Mine of Grass Valley one of the most productive and best known gold mines in California. This pursuit was in keeping with his lifelong love of history, especially as it related to his active association as a docent at the Empire Mine State Historic Park.

Born in East Liverpool, Ohio, on July 4, 1908, Mr. Steinfeld attended public schools in Pittsburgh, Pennsylvania, as well as the University of Pittsburgh. He received a degree in chemistry from Newark Technical College in Newark, New Jersey, and later in life studied mathematics—which he described as the "only pure science"—at the University of California, Los Angeles.

After his graduation from Newark Tech, he was employed as a chemist by the E. I. Du Pont Nemours Company in Newark, where he worked in the paint and pigment division for approximately 12 years. While living in Newark, he met his future wife, Ann Morgan, in the New York public library, and they were married on December 31, 1938.

In 1952, they moved to the Los Angeles area, where he was sent to open a pharmaceutical laboratory in Torrance for the Strong Cobb Company of Cleveland, Ohio. Two years later, Mr. Steinfeld accepted an administrative position with Automics International in Canoga Park, where he remained until his retirement. In February 1977, Charles and Ann moved to Grass Valley. A few months later, they both became docents at the Empire Mine State Historic Park, which had become a California state park only two years earlier.

Unfortunately, Mr. Steinfeld did not live to see his work published. He died in Grass Valley on August 5, 1994, at the age of 86. He is survived by his wife, Ann, who is still an extremely active docent at the Empire Mine SHP, in charge of coordinating tours and setting up luncheons at the Clubhouse. Her group of ladies who serve the traditional Cornish pastie luncheons for specially arranged tours have become fondly known as "Ann's Angels."

The Empire Mine Park Association is pleased and proud to present Mr. Steinfeld's work on the Bourns to the public. With deep respect, this book is dedicated to his memory.

Table of Contents

Introduction

The important role played by the W. B. Bourn family in the history of California is not generally well known. Usually, the family is heard of only in connection with "Filoli," the Bourn mansion in San Mateo County, or the Bourn "Cottage" at the Empire Mine State Historic Park in Grass Valley. The Bourn family is introduced on guided tours at these sites as their builders and first occupants, and the Bourn family wealth is usually noted as having been derived from ownership of the famous Empire Gold Mine. There is little mention of any family roles in the political, financial, or social life of the time.

This somewhat narrow interpretation of Bourn history has not gone unnoticed by visitors to these properties, and the question "Is there anything more available on the Bourns?" is frequently asked at the conclusion of a guided tour. To remedy this oversight, this book hopes to present a composite of biographical and anecdotal information on the Bourns and the related Chase, Starr, and Vincent families.

At the start of this project, the author had the good fortune to meet with Mr. and Mrs. F. Bourn Hayne of Napa County, where it became apparent that there was indeed "something more" to be told about the Bourns. Frances Bourn Hayne, a nephew of William Bowers Bourn, Jr., revealed a vast horde of family records, diaries, letters, etc., that he had gathered for 50 or 60 of his 77 years (at that time).

The sheer volume of the material was awesome. There were trunks, five-drawer file cabinets, bookshelves, and boxes, all overflowing. It was obvious that a year or more would be required to just give the material a cursory reading. To organize and cross-reference the information would require several more years. Mr. Hayne talked at some length about his hope for a definitive family history, and regretted that he, personally, had not pursued this objective in his younger and more robust years. Although he did not live to see the realization of this dream, Mr. Hayne did have the gratification of knowing that his interests in preserving the family history were shared by other members of his family, and that a respected author had been engaged, under contract, and was actively pursuing the development of a family history. As a result of this contract, access to the Bourn family files was restricted at the time this publication was being prepared. Although this restriction has since been lifted, the manuscript for this book was already in its final form and, therefore, did not have the benefit of the large mass of Bourn history that is now available to future historians.

The author wishes to acknowledge, with deep appreciation and thanks, all who have contributed to this project. Foremost, it is Francis Bourn Hayne's memories and records that form the framework of this book. More than seven hours of taped conversation with him, during which he dredged from the depths of his astounding memory details going far back into the lives of his ancestors, were carried from his home. In addition, he supplied copious notes, letters, and other important data on Bourn family members.

In addition, Mr. Hayne' s gracious wife, Ellen Hamilton Hayne, joined the discussions from time to time, clarifying points of history and jogging her husband's memory when he grew tired from long hours of conversation. Mrs. Hayne contributed immeasurably to the accuracy of the writer's notes. Mr. Hayne's daughter, Sarah Simpson, also provided friendly support, and it was her enthusiasm—after reading an early draft of this manuscript—that encouraged further work on the project.

The author also owes a great debt to Mr. A W. B. Vincent, grandson of William Bowers Bourn, Jr. Mr. Vincent read the manuscript twice, once in a rough, half-finished form, and again in its final draft. He corrected several serious errors in the text, and has been generally encouraging.

Some years past, Mrs. Peter Gallagher (Timmy) of Woodside, California, published a comprehensive history of the Filoli estate, its owners, architects, and builders. Both Mrs. Gallagher and this writer derived much of our information from the same sources—the Bourn family members mentioned above. I have not hesitated to borrow as necessary from Mrs. Gallagher's definitive work, "Filoli," for this work on the Bourn family.

Others to whom profound thanks is owed include Mrs. Nancy N. Levensaler, granddaughter of Mr. and Mrs. George Starr, who provided much useful information on her grandparents. The thoughtfulness and generosity of Dr. and Mrs. Norman Rothenberg, in sharing literature and photographs acquired during their travels in Ireland and England, are acknowledged with sincere thanks.

The author also is deeply indebted to Mr. and Mrs. Henry Bergtholdt, research docents at the Empire Mine. They and their co-workers have been tireless in the difficult reading of back issues of the Grass Valley/Nevada City *Union* going back to the 1860s. Their transcripts relating to mine owners and managers of the Empire and North Star Mines have been invaluable in developing and organizing this work.

Of special note is the writer's obligation to Mr. Hall Warren of Mill Valley, California. With a Master's Degree in history, he has taught at various California schools. He is a freelance writer and extensive traveler. Mr. Warren became interested in William Bowers Bourn, Jr., almost by accident. At various unrelated times over the past seven years, he has visited Filoli at Woodside, Christian Brothers at Napa, Muckross in Ireland, and the Bourn home on Webster Street in San Francisco. It was only at a fortuitous visit to the Empire Mine State Park in Grass Valley that he realized that the Bourn name was connected to all of these properties.

Mr. Warren offered his help on this manuscript in any capacity needed, and joint plans were formed very quickly. An accumulation of miscellaneous notes was turned over to him, and he proceeded to do essential research at Stanford University, the University of California, Berkeley, and other Bay Area libraries, from which the character of W. B. Bourn, Jr., was developed in significant detail. As a consequence, Mr. Warren authored those chapters on Mr. Bourn as an adult, which elegantly portray him against the backdrop of his historic times. The author is deeply grateful to Mr. Warren for his interest and assistance.

It is the writer's hope that this collection of anecdotes and memoirs will find its way into the hands of those who, like the author, find California history a colorful and exciting panorama.

The Bourn family was a strong force for good throughout California's Golden Age. They left an enduring mark on San Francisco through their contributions to the economic, social, architectural, cultural, and religious life of the community. That same force for good continues into the present day.

Charles Steinfeld

Old Irish Ballad

by
Mary Carton

An American landed on Erin's Green Isle
He gazed at Killarney with
rapturous smile.
"How can I buy it?" he said
to his guide.
"I'll tell you how," with a smile
he replied.

"How can you buy all the stars in the sky;
How can you buy two blue Irish eyes?
How can you purchase a fond
mother's sighs?
How can you buy Killarney?

"Nature bestowed all her gifts
with a smile—
The emerald, the shamrock, the Blarney.
When you can buy all those
wonderful things,
Then you can buy Killarney!'

Courtesy: Bob Paine

~ 1 ~

AMERICAN BEGINNINGS

The American branch of the Bourn family may be traced to the England of 1629, when a group of Puritans met to discuss pressing problems of the church and state. At that time, resistance to the Church of England had become a political as well as a religious issue.

King Charles I was viciously persecuting all who did not embrace the doctrines of the state church. When Commons denied him further funds for his extravagant war with Spain, he resorted to forced loans, billeting troops in civilian households, and heavy assessments on landowners, particularly those of Calvinist and Puritan leanings. Resistance to such high-handed measures resulted in imprisonment and martial law.

Businesses began to fail. The cost of living rose drastically, and many Puritans of wealth and long-standing influence realized that their future and safety, as well as their religious freedom, were in grave danger.

These were the grim realities on the agenda facing the Puritans in their 1629 meeting. With nothing but a dismal future ahead of them in their homeland, they began to consider the merits of founding a new colony in Massachusetts. For guidance and leadership, they turned to John Winthrop, a prominent churchman and successful attorney. If he would agree to lead them to New England and serve as governor of their proposed new commonwealth, they would guarantee to enlist a sizable group of influential citizens in the venture.

There was no difficulty in obtaining a grant. King Charles regarded ail Puritans, and other nonconformants, as a threat to the tranquility of his reign and was only too happy to see them relocate in large numbers to the wilds of New England. Furthermore, the monarch was anxious to encourage colonization as a means of neutralizing the expansion plans of the French, Dutch, and Spanish. Official details moved swiftly and a Charter for "The Governor and Company of Massachusetts Bay in New England" was signed March 4, 1629. Early the following year, the Bay Company sailed for America with John Winthrop as Governor.

There were several unusual aspects to this venture. With a fleet of 11 vessels, carrying almost a thousand men, women, and children, it was the largest expedition of English emigrants to the new world up to that time. Although the majority were Puritans, a number of other faiths were represented in the Company. There were more men of affluence and education than in previous expeditions, including lawyers, physicians, religious leaders, men of business and finance, plus a large complement of country squires, craftsmen, and artisans.[1]

The expedition also possessed unique political aspects. Among the King's normal policies on colonial grants was a stipulation that only one or two members of the governing body of the newly formed Company were permitted to emigrate with each expedition. This gave the King and Parliament the opportunity to exert direct pressure, when deemed necessary, on the majority of the Company's ruling body. A second stipulation required the Company Charter—the document itself—to be retained in the archives of the British government. This provided the King with the capability of canceling the charter, should there be any resistance to state authority after the Company reached its new home in America.

Contrary to these policies, the entire governing body of the Company was on board the ship *Arabella* with Winthrop. Winthrop also carried with him the Company Charter, something that would never have been sanctioned by the Crown. Such strategies were designed to give the leaders of the expedition autonomy over the future of the Company. They would, in effect, be a new political entity with only nominal ties to the English Crown.[2]

In general, they were strong men, displaying a startling independence of the mother country, an independence tinged with defiance even before they weighed anchor for New England. Men of bold vision, shrewd, enterprising, with diverse skills and backgrounds, they were anxious to test themselves against a raw frontier.

Enter Jared Bourn

One of the men in this party was Jared Bourn, an indentured servant, who was destined to become the ancestor of the New England family of that name. Jared settled in the Boston area and apparently prospered in the new community. No information is available on his marriage, but 25 years later, around 1655, he and his family moved to Somerset in Southern Massachusetts, just across the Tauton River from Aquidneck, Rhode Island. This probably occurred during one of the periods when many families moved from the Bay Colony to find greater freedom of religion and self-expression in the more liberal colonies of Rhode Island and Connecticut. For the next 150 years, succeeding generations of Bourns spread throughout the surrounding New England areas.

Within a generation or so, the Bourn family entered the shipbuilding business and became closely associated with a family named Bowers, who were shipbuilders and operators of a fleet of cargo vessels. Branches of both families became quite wealthy from their shipping interests, but, except for a few family legends handed down, there are no records of their achievements.

A breath of scandal always seems more intriguing than the misfortunes and triumphs of everyday living, however, and memories of family improprieties linger through the years, sometimes assuming almost a mythical or legendary character. Thus, an odd story that has survived through the years concerns a certain black sheep of the Bowers family in colonial days, as told by F. Bourn Hayne:

> When I entered Harvard in 1922, Mother, who stayed at a boarding house in Cambridge to help me get settled, took me down to Newport, R. I., to meet Augustus O. Bourn, a

former state senator of Rhode Island.[3] He was about 90 then. He was in the rubber business—that was good rubber, not what they are putting out today. He was interested in family history and had developed a genealogy of the Bourn family. Well, there was an old portrait, a wedding portrait, of a very beautiful Boston girl and her husband, who was a member of the wealthy Bowers family. It was done by a distinguished artist, Copley, I think.[4]

I don't think he was very handsome, but she was exquisite. I don't know why she ever married him. However, they seemed to get along well, and eventually had a son whom they adored and spoiled in every possible way.

Overindulged, pampered, and coddled, especially by his mother, the boy grew into a greedy, selfish adult—both dissolute and dishonest. This scoundrel, by bribing his father's accountant, managed to get away, periodically, with large sums of his father's money. He felt secure because his father had developed poor eyesight and failed to detect many discrepancies in his account books. So, the son purloined a lot of his father's money through the unfaithful accountant. In the eyes of his parents, however, he could do no wrong and was the light of their lives. Then, the old man sent his son up to Boston as his personal representative to dispose of a cargo of tea which had arrived on one of his ships, now at anchor in Boston Harbor. The young fellow went up, and on his first night in town gambled and lost the entire cargo of tea. Then, the next day, the damned fool returned to the gaming tables and lost the ship.

He was kind of afraid to go home, because Pop would be quite annoyed, so he took off on a coastwise trader, got down to New Orleans, and worked his way up the Mississippi. There, he got involved with an Indian maiden and they had a child.

Some time later, a trapper, who had heard the story from the scalawag, came East and went to Rhode Island to report to the family about their son. When mother and father got word their son was still alive, they sent money with a traveler, or scout, down the Mississippi to bring the boy back home.

By that time, Sonny was getting tired of the tribe's corn and jerked meat diet, or whatever, so, leaving the Indian Princess and the child—I don't know whether it was a boy or girl—the prodigal son returned home.

By and by, the Indian maiden began to pine away, so she put the papoose on her back and turned up in Rhode Island, but the derelict husband would have nothing to do with her. However, money mends many broken hearts, so the Indian maiden and child returned to their tribe with plenty wampum and were never heard of again.

And Sonny? You guessed it, resumed his endearing place in mama's heart.

William Bowers Bourn I

Some 183 years after Jared Bourn landed in the New England wilderness, William Bowers Bourn I, son of Francis Bourn and Mary Bowers, was born in Somerset, Massachusetts, on June 2, 1813.

Those years saw the frontier move a thousand miles west, leaving behind prosperous farms, thriving cities, universities, and cathedrals of great distinction. More important, they saw the harshly ruled bible commonwealth of the Massachusetts Bay Colony become a member state of a new nation. The American Revolution had come and gone. Ties with England had been severed, and the United States of America—a new experiment in democracy—was taking a prominent place among the nations of the world.

This fledgling country, in which W. B. Bourn grew to maturity, was a totally different environment from that of his ancestor, Jared. Survival, in an uncultivated, undeveloped land, was replaced with new challenges intrinsic to a new government, a new democratic society, and to burgeoning cities. It still was a world of manifold opportunity for those with imagination, ingenuity, and courage, and those qualities had not diminished in the six generations since Jared Bourn set foot on New England soil.

W. B. Bourn began his working career as a merchant, and while still a young man, also engaged in the banking business.

At some point in the 1840s, he entered into a partnership with a Captain George Chase in a marine shipping and trading company. Captain Chase, like Bourn, was a descendant of British immigrants. Born in Martha's Vineyard in 1803, he inherited—also like Bourn—a vigorous, forceful, pioneering spirit. He and his wife, the former Mary Elmes of Wiscasset, Maine, had six daughters and two sons.

With a small fleet of ships, Chase and Bourn traded in lumber, sugar, rice, cotton, molasses, and other staples, with the West Indies, Cuba, New Orleans, and ports up and down the Eastern seaboard. There is no evidence that the company ever engaged in slave trading.

With their common Yankee background, and with Bourn's shrewd business sense combined with Chase's experience in maritime commerce, they had an excellent team and worked together very effectively.

NOTES AND REFERENCES

1. Bailey, Thomas A *The American Pageant, A Histoty of the Republic,* Fourth Edition, Vol. I., D. C. Heath and Company, pp 25-26.
2. *Ibid*
3. Augustus O. Bourn, a Republican State Senator, and wealthy rubber manufacturer from Bristol, Rhode Island, was a first cousin of W. B. Bourn I. In addition to developing a Bourn family genealogy, he was the family member who saved all the letters Aunt Annie Ingalls wrote to her husband while she visited her brother in California. He also saved the letters of Sarah Bourn to her brothers-in-law, Benjamin and Phillip Bourn, describing her early days in California, which is how they came into the possession of F. Bourn Hayne.
4. From a time standpoint, the portrait certainly could have been painted by John Singleton Copley (1738-1815), a renowned American portrait painter.

~ 2 ~

THE FAR WEST BECKONS

News traveled slowly. It was almost seven months after Marshall picked up his first nugget that word of the gold strike began to appear in eastern newspapers, but most of the people were skeptical and unmoved.

Doubts faded somewhat when President Polk made public a report from his military governor in California citing the frantic activity along the American, Feather, and other Northern California rivers and streams. The report was accompanied by the most compelling evidence—a pouch containing 230 ounces of gold. However, it was the glowing letters from settlers to their families and friends back East that finally aroused the interest of even the most skeptical, and the race West began in earnest.

Captain Chase had become interested in California prior to the gold strike. Following the overthrow of Mexican rule in 1846 by General John C. Fremont and the creation of the Bear Flag Republic, Chase began to watch developments taking place along the Pacific Coast. When the Treaty of Guadalupe-Hidalgo was signed in early 1848, giving the United States title to Texas and the vast territory stretching to the Pacific, Chase realized it would only be a short time before California would be admitted to the Union. It was a virtual certainty that commerce between the Hawaiian Islands, the Pacific Coast, and the East would skyrocket, and the lure of the new western frontier could no longer be resisted. Outfitting the *Robert Fulton,* organizing a company of men who wanted passage to the gold fields, and leaving his partner Bourn to manage affairs at home, Chase set course for Cape Horn and San Francisco.

The Captain's ship was wrecked off the Falkland Islands, but he managed to obtain another vessel from the British and proceeded on course. In late 1849, he dropped anchor inside San Francisco Bay. No passengers were lost on the voyage, but Captain Chase's bible was damaged by seawater and the family entries in the front of the bible are no longer legible.

Opportunities for New Ventures

It took Chase only a few days to realize he had correctly assessed the prospects for a profitable shipping operation in the bustling new community. The opportunities were obvious.

As so frequently happened in the early days of the gold fever, his crew promptly deserted, heading for the placer mines along the American River. Chase, however, gambling on his ability to keep future crews aboard ship, lost no time establishing a West Coast base of operations. Almost immediately, however, he ran into difficulty. His shipping agent became seriously ill and other problems arose that were too much for one man to handle. Thus, within a few months after arriving in San Francisco, he was forced to send word to his partner, Bourn, to come to San Francisco as quickly as possible. It was not the best timing for such a request. For several years, Bourn had been courting Captain Chase's eldest daughter, Sarah, and just a few months after the Captain had set out for California, Sarah Esther Chase and W. B. Bourn were married in the Church of the Holy Trinity, Brooklyn, New York, on July 24, 1849. The Captain's urgent plea for help arrived a few months later in early 1850.

To ignore the urgent request of his partner (now his father-in-law) was simply unthinkable. Leaving his bride in Yonkers, New York, with Georgina, her married sister, Bourn took off. By taking the overland route across the Isthmus of Panama, and then sailing up the Pacific Coast, he arrived in San Francisco in late 1850, just after California had become the 31st state in the Union.

He intended to stay only six months, which, counting a three-month trip each way, would have kept him away from Sarah for a year. But the new land and the exciting opportunities open on all sides forced Bourn to pause and re examine his long-range goals.

He found himself in a flourishing community that could someday be one of the world's major seaports. The new State of California, with its natural beauty, fertility, and vast mineral resources, was certain to become a powerful force in the Union. Just as the eastern frontier of 1630 had presented an irresistible challenge to his ancestor Jared Bourn, the far west of 1850 held the promise of a future limited only by a man's imagination.

Bourn wanted a part in that future, along with a share of the action in this exciting city teaming with frenzied miners and prospectors passing to and from the diggings along the northern rivers. Everywhere, there were men who had an utmost need of goods and services of every description, and they were willing to pay exorbitant sums to obtain them.

It was a place of infinite opportunity, but it also was a wild, primitive frontier town. Consequently, more than three years passed before Bourn felt it was safe to send for Sarah.

The surroundings were extremely crude. The new section of town was a tent city with few substantial buildings. In the rainy season, mud ran so deep in the streets that they were impassable. Open sewers were a constant invitation to disease. In dry weather, the air was filled with choking dust, churned up by horses and wagons. The human environment, however, was far more dangerous than the physical.

Even more than the gold camps of the interior, San Francisco had attracted the worst elements of society. Night and day, robbers and ruffians roamed through the streets that were lined with saloons, brothels, and gambling houses. Drunken rioting was a continuing part of the daily scene. By 1851, murders, stabbings, cruel assaults, and arson reached a level the authorities were powerless to curb. Believing their only hope of survival was to take the law into their own hands, the terrified citizens organized the original "Committee of Thirteen." Hundreds joined the movement and the day of the Vigilantes arrived. After a hanging and the eviction of a number of the worst offenders under threat of death if they returned to the city, things quieted down—relatively speaking—and the city began to improve. Before long, permanent buildings gradually started to appear.

Sarah Comes to San Francisco

With San Francisco becoming more civilized, Bourn finally wrote to Sarah to start her journey West. Traveling by ship to Nicaragua, by foot and mule-team across the Nicaraguan Isthmus and again by ship, Sarah made it to San Francisco on May 4, 1854.

It was a forbiddingly difficult journey, even for a strong man, and it was an emotional meeting when Sarah was greeted at the ship by her husband, her father, and the rest of her family, which Captain Chase had brought to San Francisco about two years previously.

One of Sarah's first letters back East was to her brothers-in-law, Phillip and Benjamin Bourn. The letter described the rigors of the trip and the subsequent pleasure of being reunited with her husband and family. During her first years in San Francisco, Sarah continued this correspondence, filling it with her impressions of life in her new home in the West.

~ 3 ~

AT HOME IN SAN FRANCISCO

It had been four long years since Bourn had held his bride in his arms, and it was a second honeymoon for both. For a short time, Sarah and William lived with the Chase family, then they moved to a house on Bryant Street for a brief period, and then to another house at Third and Brannon Streets.

During those years, the character of San Francisco changed dramatically. Permanent buildings replaced shacks. Wealthy residential areas with elegant homes were developed. Schools and churches were built. Many streets were still a sea of mud in wet weather, and robbery and mayhem yielded to law and order grudgingly, but with law enforcement once more in the hands of government authorities, the responsible citizens began to plan the future metropolis.

With Sarah's loving support, W. B. Bourn flung himself energetically into the business world just beginning to take shape in the new community.

He was a vigorous, hard-driving man, never content with just one iron in the business fire. In addition to his responsibilities with the shipping firm, Bourn entered the commission brokerage business at an opportune time of growing demand. He resumed his banking interest, became a director and then president of the Farmer's Fund Insurance Company, entered actively into trading on the Mercantile Stock Exchange, and began to invest in diverse mining ventures, including the Empire Mine in Grass Valley.

While Bourn was thus laying the groundwork for a large fortune, Sarah was rearing a family of two sons and four daughters:

Mary Champney Bourn (later called Maye), born in San Francisco, February 11, 1855; married James Ellis Tucker in 1895, had no children, and died in San Francisco, November 2, 1947, at the age of 82.

William Bowers Bourn, Jr., born in San Francisco, May 31, 1857; married Agnes Moody in Yonkers, New York, 1881; they had two children, a son who died in infancy and a daughter, Maud Eloise Chase Bourn. William Bourn, Jr., died at Filoli, July 5, 1936, age 79; Agnes Moody was born in New York City, November 1860, and died at Filoli, January 1936, age 76.

Sarah Frances Bourn (called Zaidee), born in San Francisco, June 5, 1859; married Horace Moody of New York, brother of Agnes Moody Bourn; had no children; died San Francisco October 1898, age 39.

Frank Washington Bourn, born June 20, 1861; died San Francisco in January 1872 at the age of 11 from injuries sustained when he fell from a stone wall.

Ida Hoxie Bourn, born Brooklyn, New York, April 2, 1864; never married; died San Francisco, December 1951, age 87.

Maud Eloise Chase Bourn, born San Francisco, November 15, 1867; married William Alston Hayne II, December 1899; two sons, William Alston Bourn Hayne and Francis Bourn Hayne; died June 1, 1948, age 81.

Young William Bourn, Jr., attended Bates School in San Francisco, and a military school in Benicia, known as the College of St. Augustine. The girls and Frankie all attended local San Francisco schools. After finishing their school work, the leisure time was occupied by music lessons for the girls, and that marvelous new game "baseball" for the boys.

San Francisco matured rapidly, and as it grew in wealth and importance, it attracted lawyers, doctors, ministers, architects, and educators. As the years passed, the journey between East and West decreased, from a doubtful three months to less than four weeks. Travel became less arduous, as steam trains across the Isthmus replaced mule trains and foot-trails. Ships were more comfortable and their food was more palatable.

As a result of these advances, visitations between families became more and more feasible, and Sarah celebrated the 10th anniversary of her arrival in San Francisco while on a visit to the family back East. Sarah gave birth to her fifth child, Ida Hoxie Bourn, while in Brooklyn on that visit in 1864.

Just a year later, the Bourns entertained Mr. Boum's sister, Mrs. Zebediah Hanna Ingalls (called "Annie") and her son, William Bowers Bourn Ingalls (called "Willie"). Willie was born in Brooklyn on October 11, 1853; therefore he was only 12 years old at the time of his visit to San Francisco.

Willie's Diary

Willie faithfully kept a diary throughout the entire trip, which was published many years later in the *California Historical Society Quarterly.* As seen through the eyes of a young boy, it is a colorful, intimate account of cross-country travel and of life in California at the close of the Civil War. His intuitive awareness of aesthetic values in life around him, his skills of observation and description are remarkable.

The diary merits the attention of those interested in California history of 1865-66, and is presented here with permission of the California Historical Society. (Willie's original spelling, grammar, and style have been retained.)

Steamship New York

Oct. 2nd, 1865: We left our good old home at State St. (Brooklyn, N.Y.) at nine o'clock AM. for the steamer bound for California. Upon arriving at the steamer, all was noise and confusion—we went directly to our state room and unpacked our carpet bag and got ready for the night. We went out into the cabin to see our friends who kindly came to bid us goodbye. After we got off, mother and I went upon deck and staid up till we passed Sandyhook. At Ft. Lafayette they fired three or four guns and I could see the light from the guns.

Oct. 3: I was sick all day and did not go to the table.

Oct. 7: When I got up this morning I was greatly surprised to see land; it proved to be one of the Bahama Islands. As we passed by, a boat came out to our ship to get the mail. I saw on the Island some houses and great piles of salt, and it glistened in the sun like diamonds. We passed Cuba, it is green and mountainous, I saw a ship near there. The weather is very hot and I am glad to get in my thin clothes.

Oct. 10: We arrived at Aspinwall (former name for Colon) at six in the morning. I got ashore as quick as I could, and went up in the main street with the rest of the passengers. I saw some cocoanut trees and the nuts growing. I went to mother for some money and I went back to get some bananas and took them back to mother … mother and I then went to the Consuls office and left our things; then we went out and mother bought a large basket filled it with bananas of red and yellow and she bought two oranges, then we went back to the consul office and staid there until the cars went and then we got in the cars to cross the Isthmus. It was quite a change to get in the cars after being so long on the ship. I saw some oranges and bananas growing and I saw a parrot flying about. We stopped at Cruces that is halfway across the Isthmus. The natives brought their fruits to sell. They had some native bread for sale and oranges, limes, native candy and eggs. The native candy is cocoanut candy and maple sugar. I forgot to mention we followed Chagres river as far as Cruces. The bridge over the river is made of iron. From Cruces to Panama it is very much prettier than from Aspenwall to Cruces. We arrived at Panama about five in the afternoon. We went right to the transport boat as the passengers are not allowed to stop there. It was very warm and so crowded that we had no place to put our things. I was very glad to get to the steamer. Panama is an ancient looking place; there are two or three ruins in it and it looked very pretty as we left the harbor. We had to go about three miles to the steamer *Colorado.* It is a very large steamship, I should think about twice as long as the *New York,* but I do not think they have as good a table as on board the *New York.* At the childrens table it is miserable, but fortunately mother got me at the first table.

Tuesday, October 24: We arrived Tuesday morning at San Francisco about quarter past eleven. As we went in the harbor, we fired a gun off Megges Wharf—it made us jump like everything. At the entrance there is a rock with a hole through it and if you get in such a position it will look just like a key hole and they say it is the lock to the Golden Gate. And at the entrance there is a very strong fort—they say it is the strongest fort in the world.[1] San Francisco is a very different city than what I had thought, it is very hilly, I thought it was rather level. Before we got in there was a great crowd, we could see the carriages waiting for the passengers. Before we were barely in there were hacks on board to see if they could get somebody to get their carriages. There was great confusion. Mother staid by her stateroom so as to watch it. I went out on deck to see if I could see Uncle William. But I went in to sit down with mother and I saw somebody look like him and sure enough that was him. We waited until the crowd got off the steamer and then we went to the carriage and rode up to the house. As soon as we got there Mary came rushing out the door to meet us.

Wednesday, Oct. 25: This morning Aunt Sarah, mother and I are going out to see the city. Aunt Sarah has some shopping to do and we are going with her. She took us up on one of the princible streets and she showed us some of the finest houses I ever saw. I think they go ahead of Mr. Hunt's houses and she showed us the house that Uncle William thought of buying; it is upon a hill so that you have a full view of the city.[2]… We went over to see Mrs. Chase—she is a great deal better than what she was two or three days ago for they thought that she was dying.

Christmas Day, 1865: It is very pleasant this morning. All of us got up early this morning to see our presents; we had a Christmas tree and Mary and I dressed it all alone. I went to Grace Cathedral, it is a fine church.... When church was out I came home to get ready to go to Mrs. Tibbey's to take dinner[3]: the whole of Mrs. Tibbey's family were there. We had a very nice time indeed. In the evening she had a party; there were about fifty present. I enjoyed it very much in dancing and watching the others dance. Uncle William being absent for about a week came in very unexpectedly about 12 o'clock. He was introduced to all the ladies and kissed about all of them.

Feb. 23rd: This morning I went over to the New Almaden Mines—they are quicksilver mines and they are the richest in this country ... before we go down in the mine there is a tunnel that goes into the ground eight hundred feet before it goes down. At the end of the tunnel there is an engine and that works, and brings the silver out of the mine.... I went down to the bottom of the mine and I knocked some pieces off and the men said they were very rich.... We rode out of the canyon and stopped by the roadside and ate our lunch under a bay tree and it was all in blossom and it was beautiful to the hills all about us.

April 7: This morning we started for Napa at eleven o'clock—we took a boat and landed at Suscol about half past two and we took a train which took us to Napa—it was very warm there and I was glad to get out of the place. Then Uncle William got a carriage and we started for the White Sulphur Springs; we did not arrive at the springs until about eight o'clock. We had to cross two streams and at one place the driver had to get out to feel the way before crossing as there was quick sand there and mother says she will never forget it.

April 8: This morning I got up very early to gather some wild flowers for mother and I went up on the high hills to gather them. The hotel is situated right in the canyon on the side of a mountain and there is a stream of water running by the side of the house and I went bullfrogging and I caught five of them, and I had them cooked for supper and they thought they were very nice.

There are five different sulphur springs and they are very much stronger than those of San Jose in the taste of sulphur and the water was yellow with the sulphur and the water was warmer.

April 9: This morning about eight o'clock we started for the Calistoga Springs which is about ten miles from the White Sulphur Springs and we arrived there about eleven o'clock and it is very pleasant there, only it is lacking for the want of trees. The springs are boiling and it will boil an egg in three minutes and we boiled one and it tasted as well as though it had been boiled by the fire.... There is a swimming bath up there and it is warm, and Uncle William, Willie,[4] and myself all went in bathing. Uncle William and I swam out in the deep part and dove down and had some fun. I must tell you about Willie; he had been boasting of his swimming, saying that he could swim better than I, so I let him keep on, so when I got in I just swam out to see if he would follow me. But instead of following me he kept hold of the rope, standing there and looking at me and he could not swim a stoke, but the funniest of it was he said he could swim in the Hudson R. but he could not swim in the bath....

Steamship Northern Light (to New York)

July 1: This morning it is very smooth, it is just like glass and is nice and cool. There are seven sails in sight and two steamers—I have kept count of how many vessels we have passed on this ride. I have counted 43 and I don't know how many we passed in the night. We have gone since twelve yesterday to twelve today 237 miles and we have 109 miles to go before we get into New York.

Willie was not the only writer in the family. His mother, Annie Ingalls, sent off some 30 letters to her husband during their seven-month visit with the Bourns. It is remarkable that even though the Confederate surrender at Appomattox had occurred just six months prior to their visit, there is no mention of the Civil War in any of her letters.

Annie Ingalls' California Impressions

California apparently captivated Annie as much as it did her young son. Her letters are filled with descriptions of California scenery and life and events in San Francisco, running the gamut from shopping sprees, the fun the boys were having throwing clay balls at the barn door, on up the scale to the vitally serious problems in social etiquette posed by the arrival of a new Episcopal Bishop and his young wife. Should, for example, Mrs. Bourn promptly make a call on the wife of the new Bishop to welcome her to the parish, or should she wait for the Bishop's wife to call upon the Bourns? This problem alone provided hours of soul-searching debate.

Annie's letters also include an adult's perspective of some of the events covered in Willie's diary. Among these were the inspection trip to the New Almaden Mercury Mines and the excursion to White Sulphur Springs and Calistoga.

In writing about the latter event, Annie expands on Willie's account to note that the train tickets from Suscol to Napa, a distance of five miles, were one dollar a person, which everyone thought was scandalous for such a short trip. She observed that White Sulphur Springs was founded in 1852 and was one of the first resorts of its kind in California.

Annie also concluded that the men in and around San Francisco must have been very heavy drinkers, because so many of them drank the foul-tasting sulphur water in Calistoga as a sort of cleansing therapy, similar to the English practice with the spring water at Bath, England. She decided everyone believed that the nastier tasting a medicine was, the better it was for you. She also was somewhat amazed by the luxurious accommodations and special conveniences provided to these men; they even had a special telegraph line to the baths so they could keep abreast of stock market trends.

Finally, while the boys were in swimming at Calistoga, as told by Willie, Annie and sister-in-law Sarah indulged in a mud bath, which was very warm and comfortable, and they could talk to each other over a board partition that capably preserved their Victorian modesty.

Captain Chase did not escape Annie's sharp sight, either. She mentions him in several letters. The good Captain, who loved parties, was a frequent guest when the Bourns entertained. With a little too much wine aboard, dignity apparently deteriorated and he would go around kissing all the ladies present. This seemed to be just a bit scandalous and embarrassing to this lady from New York.

It was a less complex world. Her letters were simply addressed to Mr. Z Ingalls, Brooklyn, N. Y. There was no street address, and zip codes were almost a hundred years in the future, but every letter was faithfully delivered.

NOTES AND REFERENCES

1. This is a reference to Ft. Point, which was built at the beginning of the Civil War to protect the San Francisco Bay from Confederate ships. It is directly under the south end of the Golden Gate bridge, and now is a National Monument.
2. Uncle William did buy the house at 1105 Taylor Street on Nob Hill, just North of the present Grace Cathedral.
3. Mrs. Edney Stagg Tibbey was the former Emily Chase, Mrs. Bourn's sister. Her husband was receiving teller at the Bank of California, organized by William C. Ralston. The Tibbey residence was at 923 Howard Street. Mr. Tibbey played an interesting role in the early history of the Bank of California. See *Ralston's Ring,* by Lyman, George D., Scribner, 1937.
4. The "Willie" referred to in Willie Ingalls' diary account of swimming at Calistoga apparently was William B. Bourn, Jr., who also was called by that name as a child and would have been eight years old in 1865.

~ 4 ~

MADROÑO—FROM TOWN TO COUNTRY

While Sarah was busy with her family, her husband continued to diversify his business interests. Investing more and more heavily in mining stocks, in 1896 he acquired controlling interest in the Empire Mine at Grass Valley. He was also a director and president of the Imperial Silver Mine of Virginia City and the original Hidden Treasure Mine of White County, Nevada. From that time on, a large part of the family fortune was derived from mining.

Shortly after the birth of Maud, their last child, Mr. Bourn presented Sarah with a country home at St. Helena in the Napa Valley. Known as Madroño, the estate originally consisted of a main residence with barns and outbuildings, along with two comfortable farm houses known, respectively, as the "Red Cottage" and the "White Cottage." There were 60 acres of fine vineyards stretching south from the main house. Within a short time, Bourn added to this by purchasing from Sam Brannon (of San Francisco fame) another 60 acres of vineyards, adjacent to and lying north of the main house.

Although San Francisco had made remarkable strides, and the Bourn house at 1105 Taylor Street was an elegant and comfortable residence, the city retained many vestiges of its brawling, frontier origins, and Sarah welcomed her husband's gift in St. Helena as an opportunity to escape with her children to the quiet, pastoral charm of the Napa Valley.

Mrs. Bourn had the original Madroño house torn down and replaced with a very palatial home, which utilized large quantities of decorative stone work and boasted two imposing towers in the Victorian style of the period.

In 1872, 11-year-old Frank died of injuries sustained when he fell from a high stone wall at their home on Taylor Street. Following this tragedy, Mrs. Bourn and her daughters spent more and more time at Madroño, with Mr. Bourn and William, Jr., coming up from San Francisco as frequently as business or school would permit.

Bourn Luck

Young Bourn, Jr., was particularly fond of performing card tricks for the girls in a room at the top of one of the towers. He was very adept at manipulating cards and could flick a card the length of the family's new living room with astonishing velocity and accuracy. His skill extended beyond card tricks, as he became an above average card player while still a youth. Later on as an adult, his frequent winning at cards with his business associates, as well as his unusual successes in investments and business deals, began to earn him a reputation among his friends around San Francisco that became known as "Bourn Luck."

Madroño was a country retreat in the best possible sense, and the family loved it. Sarah was especially fascinated with its many possibilities. In addition to her vineyards, she enjoyed experimenting with other crops and animals, including silkworms, Chinese pheasants, dairy cows, and chickens. Her Madroño bred carriage horses became well known throughout the valley.

Surrounded by her children, her vineyards, and her farm activities, the estate was a joyful sanctuary, which she shared happily and generously with two of her sisters and her aging father, Captain Chase. Sister Charlotte, who had married a William Starr in San Francisco, lived in the Red Cottage with her young son, George W. They were abandoned by the husband when George was very young. A second sister, Caroline Chase Bowers and her two daughters, Manzanita and Edith, lived in the White Cottage and cared for Captain Chase, long retired from the sea and badly crippled with arthritis in his later years.

Into this idyllic life, sorrow came all too swiftly when Sarah's sister, Charlotte Starr, died at the Red Cottage on April 11, 1874. Sarah mothered George W. Starr, keeping him busy with odd jobs around the estate. The year was to be one of continued tragedy when, on July 24, Sarah and her daughters were called back to San Francisco from Madroño to learn upon their arrival at the Taylor Street home that William B. Bourn, Sr., had died—at age 61—from what was presumed to be a self-inflicted gunshot wound. It was their 25th wedding anniversary.

W. B. B. Ingalls, the "Willie" of 1865 and nephew of the deceased, found the body. Employed for the summer as a clerk in a San Francisco insurance office, Willie had been sent to the Bourn home to tell his uncle that he was needed at the insurance office. William Bourn, Jr., arrived at the home a few minutes after his cousin had made the shocking discovery.

Mr. Bourn had not been under a doctor's care, nor was he taking any medication; however, he had been troubled with severe head pains for some time.

A coroner's jury, which convened the following day, examined only two witnesses, the 21-year-old nephew who had found the body and the 17-year-old son.[1] The jury returned a verdict of suicide. About three weeks later, a second coroner's jury was convened to hear testimony of several of Bourn's business associates who had talked with him at length the day prior to his death. They all testified that there was no evidence of depression or any other visible indication of the possibility of suicide. They also admitted that Bourn had always been rather careless in the way he handled firearms.[2]

What happened can only be a matter of conjecture, but Bourn had recently terminated the services of the Wells Fargo Company for delivering the payroll to the Empire Mine and was making this delivery personally. As a consequence, he had started to carry a revolver. The second jury, after reviewing all testimony—including that of the first hearing, changed the verdict to accidental death.

Sarah Takes Over

Following her husband's death, Sarah faced new and strange burdens. She was already busy raising and educating five children—ranging from Maude, 7, to Mary, 19, plus her ward, George Starr, 13—as well as overseeing the operations of a working vineyard. Enlisting the aid and counsel of the family attorney and a few valued friends, Sarah stepped into his void in her life with courage and determination.

Taking control of her husband's many ventures, she displayed a surprising astuteness in business matters, with extraordinary ability to reach prudent decisions. As a result, she managed to keep the life of her family flowing calmly and remarkable free of scars from their tragic loss.

In the years since Sarah had joined her husband in San Francisco, they had acquired a wide circle of friends, including many important figures in the professional and social life of the city. These friends became doubly precious as they stepped forward with heartening support and wise counsel.

One person in particular played a highly significant role in the Bourn family life—Horace F. Cutter, who was born in Boston, Massachusetts, in 1821 and lived for some 50 years in San Francisco. Cutter had gained some fame as the "Don Horatio" in Clarence King's delightful letter from Spain, subsequently published in May 1886 in the *Century Magazine* as the "Helmet of Mambrino."[3]

Cutter had close connections with important members of the Boston and Washington, D. C., society and, in San Francisco was a most welcome visitor in half a dozen houses where he was an expected guest for dinner once or twice a week.[4] A gentleman of great intellectual capacity and international interests, he was a frequent visitor at the Bourn homes in San Francisco and St. Helena. After William B. Bourn's death, he was a constant guest at Sarah's home and exerted a great influence on the Bourn family.

Francis Bourn Hayne, in later years the family historian, was convinced it was Horace Cutter's influence that led to Sarah's belief that no school in the United States could do as much for her son as would England's Cambridge University.[5]

Another strong, secondary motive might have been the belief that moving W. B. Bourn, Jr., out of the sphere of family influence—which was now all feminine—and into a new environment would broaden his cultural horizons and develop a self-reliance much needed by a young man who would eventually be faced with management of a large, diversified estate.

Shortly after completing his secondary school studies, William B. Bourn, Jr., went east to his Aunt Annie Ingalls. Following his mother's wishes, she put him on the ship for England and Cambridge University, where he would stay for three years.

Cambridge Days—Are Ye Daft Man?

Young Bourn was first admitted to Sidney Sussex College and subsequently to Downing College. It was at Cambridge that the first signs of his faculty for taking long odds and winning began to be displayed. This is best demonstrated by the following story told by A W. B. Vincent, grandson of W. B. Bourn, Jr.:

> In those days, the Cambridge barbers also acted as bookmakers, accepting and covering all bets on sporting events for the students and local citizens. In 1877, young Bourn, in need of a haircut, entered the shop of one of these barbers. Once he was seated, the barber began his usual small talk on upcoming athletic contests.
>
> "Who do you think will win the boat race between Oxford and Cambridge this weekend?" he inquired.
>
> "Oh, I don't know," responded Bourn, "I haven't thought much about it, but I suppose Cambridge."
>
> "I will be very glad to book a wager for you," said the barber, mentioning what he felt were very generous odds.
>
> Bourn was silent for a moment, and then: "What would be the odds against the two schools rowing a dead heat?"
>
> The barber laughed. "That could never happen. It is so close to the impossible I would give odds of 1,000 to 1 on such a bet."
>
> This time there was no pause. "Good," said Bourn. "I'll lay you a 10-pound wager that they row a dead heat."
>
> "Are ye daft man? That is so improbable, I hate to take your money."
>
> "Take it," said Bourn.

That was the year the impossible happened. On March 24, 1877, Oxford and Cambridge rowers crossed the finish line simultaneously, in a time of 24 minutes and 8 seconds.[6] The loss of 10,000 pounds ruined the barber and he closed his shop; however, three weeks later Bourn set him up in business again.

Legends arise out of such events, and in later years, as Bourn gave repeated evidence of his willingness to back his own judgment against recognized experts in mining and in other areas, the term "Bourn Luck" became more and more in common usage by his business associates.

At the end of his third year and the attainment of his 21st birthday, Bourn left Cambridge and returned to California to assume some part in managing the family affairs; a burden that Sarah had carried alone for nearly four years.

Today, visitors to Downing College may view the solid silver candlesticks presented to the College by Bourn in memory of his years at Cambridge.

NOTES AND REFERENCES

1. San Francisco *Alta,* July 25, 1874.
2. San Francisco *Alta, August 16, 1874.*
3. King, Clarence, *The Helmet of Mambrino,* The Book Club of California, 1938, University of California Press.
   King, also a close friend of the Bourn family, was born at Newport, Rhode Island, January 6, 1842. By the age of 30, he was an acknowledged leader in the then new and almost romantic field of geology. He was a master of expression in his scientific writing and in the world of fine literature. *The Helmet of Mambrino,* first published in the May 1866 edition of *Century Magazine* was republished in 1904 in a volume of "Memoirs" by the Century Association. W. B. Bourn, Jr., presented a copy of the 1904 edition as a gift to his wife, Agenes Moody Bourn; republished again, in a limited edition, by the Book Club of California.
4. For an excellent sketch of Horace F. Cutter, see James D. Hague's contribution in the Introduction to *The Helmet of Mambrino.* Hague was one of King's associates in the moumental Fortieth Parallel Survey, who knew well not only King, but was also a friend and admirer of Horace Cutter.

5. Horace Cutter died July 13, 1900, and is buried in the Bourn/Hayne family plot in St. Helena, close by those who valued his friendship so dearly.
6. Source: the University of Cambridge, University Archives, the University Library, West Road, Cambridge CB3, England, United Kingdom.

~ 5 ~

EARLY YEARS OF A NEW ERA

Following her husband's death, Sarah continued operations at the Empire, relying heavily on Frank Nesmith, her mine superintendent. For some years, mining in the area had been at a very depressed level, and by the time young Bourn returned to California, Grass Valley was considered to be a worked out, dying gold camp.

Many years later, George Starr described this low ebb in Grass Valley's fortunes as follows:

> There were but three mines in active operation, the Idaho, the New York Hill and the Empire, with the first in 'bonanza,' the other two in 'borrasca.'[1]

For some time, Sarah and the managers of the W. B. Bourn estate had watched the income from the Empire decline until it was barely meeting operating costs. Recognizing they must reach a decision on the future of the mine, three well known mining experts were employed to examine the workings. It was the unanimous opinion of these consultants that the Empire was worked out, and that even with richer ore, the mine had reached a level—about 1,200 feet on the dip of the ledge—considered too deep for profitable operation. On the basis of these findings, preparations were underway to abandon the property.

Once her son was back home, Sarah quickly gave him responsibility for overseeing the mine at Grass Valley and the vineyards at Madroño. Young Bourn made his personality felt immediately. After an extended visit to the mine, including an inspection of all accessible working levels, he decided to continue operations and to initiate an aggressive exploration program.

His first step was the formation of a new company, the "Original Empire Mill and Mining Company." All of the shareholders of the Empire were invited to join in the reorganized company, but all declined except one man, C. F. Fargo. For many years, the Fargo estate retained this interest, originally acquired in 1856 just two years after the mine had first been named the "Empire."

Taking over in 1879, the reorganized company began pumping operations and resumed shaft sinking. Operations were conducted under great difficulty, financially and otherwise, for this was the time when belief in California quartz mining had all but vanished. Faith and energy triumphed, however. It was not long before daily assays showed improvement, and exploration indicated the presence of large bodies of quartz with good potential.

In 1883, the Empire was producing well once again. Over the next three years, the company installed water power (at an expanse of $100,000), reconstructed and improved the surface plant, and increased the 20-stamp mill to 40 stamps. Several times in later years, the veins gave evidence of pinching out, but Bourn's faith in the mine never wavered, and in every instance that faith was richly rewarded.

In 1881, Bourn brought his 19-year-old cousin, George W. Starr, to work at the Empire. George's Aunt Sarah had raised him after his mother's death when he was about 13 years of age. She became his legal guardian and insisted that he attend St. Augustines, the same military academy that Bourn, Jr., had attended.

Also in 1881, young Bourn married Agnes Moody of Yonkers, New York. The following year, a son was born, who died shortly after birth. In 1883, a second child, Maud Eloise Chase, was born.

By this time, though he was only 26, W. B. Bourn, Jr., was beginning to move in business and financial circles with all the confidence, boldness, and acumen of his father.

Bourn Takes Control

About two miles west of the Empire was a series of claims making up the North Star Mining Company, owned for many years by a group of San Francisco investors. The North Star had produced very rich ore throughout its early years, but in 1875, at a depth of 1,200 feet, the vein was lost and, almost simultaneously, pumping problems were encountered that resulted in the mine' s closure for the next nine years.[2]

As stated previously, this was a depressed period in Nevada County. Ore quality had deteriorated generally, and there was a feeling that the best days of mining were in the past. Many of the miners had departed for Virginia City and the Comstock Lode. In a few years, however, they started to drift back to the Grass Valley area. The Comstock was going through its own depression and the miners found the heat in the lower levels of the Virginia City mines to be intolerable compared with the relatively cooler temperatures of the Nevada County mines.[3]

In 1884, Bourn was offered an option on the North Star and asked an old friend, Alexander Stoddard, who had invested in several mines in the Grass Valley area, to join him in a project to reopen the mine. That same year, Bourn employed John Hayes Hammond as consulting engineer at the Empire. Sometime previously, Hammond had inspected mining properties for Stoddard, so he was promptly pressed into service to investigate the North Star.[4]

After examination at all levels not flooded, Hammond was convinced that the faulted vein could easily be found and that it probably would develop valuable ore shoots. Based on these findings, Bourn and Stoddard purchased the mine, forming the North Star Mining Company, with Hammond as manager.

Within two years, profits from the mine were sufficient to warrant enlarging the scope of operations and justify erection of a modem 30-stamp mill. By 1886, it was the model plant in California, and over the ensuing years the North Star produced millions of dollars in dividends for its shareholders.[5] Success at the Empire, followed by the wonderful results at the North

Star, gave Grass Valley quartz mining an impetus that was felt throughout the state.

With Bourn taking greater control of the family's business interests, Sarah turned her attention to her daughters. During the early years, the girl's education was provided by tutors. Lessons were principally music, literature, and languages. By 1883, Sarah's youngest daughter, Maud, was 16, and Sarah decided they were now all of an age that she could move the family to New York and enroll them in a finishing school. This kept Sarah away from home for extended periods. Around 1886-87, she followed the accepted practice of wealthy families of adding a final touch to a girl's training by an extended tour of all the major cultural centers of Europe.

Sarah recognized her son's competence and, in general, gave him a free hand. Early in 1887, however, while abroad with the girls, she received a report from her personal business agent that included some comments about certain shenanigans going on at the mine. A letter came flying back to her son with the order: "Close the mine."

The order provides strong evidence that, even at that late date, some eight years after young Bourn reorganized the company and became its president, Sarah still played a role in management of the estate in general, and the Empire in particular. The mine was not closed, but, presumably, the letter had a significant corrective influence.

Although the terms of the senior Bourn's will are not known to this writer, it is known that each member of the immediate family was bequeathed a share in the ownership of the Empire Mine. Eventually, this arrangement irked Bourn. He did not accept gracefully the fact that on certain decisions he was obliged to consult not only with his mother, but also with four sisters who were much more interested in receiving dividends on a frequent basis than they were in problems of mine management. His older sister, Maye, in particular, was very extravagant and, consequently, always wanting more profit from the mine.

Bourn finally succeeded in buying out the interests of his mother and sisters, and, although that made life significantly easier for him in day-to-day decision making, it created a big family feud. After agreeing to the sale, the ladies put their heads together and decided they had been cheated. There was a great squabble, but Bourn stood his ground and eventually the ill feelings dissipated.

About that time, Bourn met and formed a strong friendship with James D. Hague, a distinguished engineer representing powerful eastern financial interests. Hague was roaming the western states in search of attractive investments for his clients and himself. Accepting Bourn's invitation to visit Grass Valley, Hague was given an opportunity to examine the workings and the books of both the Empire and North Star mines.

Hague Buys North Star Mine

In 1887, shortly after the new stamp mill was completed, Bourn sold a controlling interest in the North Star to Hague and his backers. Hammond resigned as manager of the North Star Mining Company and the Original Empire Mill and Mining Company.

In his autobiography, Hammond gives a good indication of the close relationship that developed between himself and the two powerful mine owners, Bourn and Hague:

> From the beginning of our association Hague, Bourn, and I proved congenial spirits. We spent our time, as Hague expressed it, with quartz by day and pints by night. The new owner of the North Star, however, although most cultivated and delightful and an accomplished engineer of wide experience, did not have a nose for a mine. This is well illustrated by what happened when Mr. and Mrs. Bourn, my wife, and I went on a vacation to the Yosemite Valley to enjoy some of the profits he made in the sale of the North Star. We had not been gone long when I was overtaken by a frantic telegram from the unhappy Hague:
>
> *Pipeline supplying power burst. Mill shut down. What do you advise?*

After deep consultation, Bourn and I dispatched the following wire to Hague:

> *Advise mending pipeline and restarting mill.*

With all seriousness Hague followed this sage, though obvious, counsel. Although endowed with brilliant qualities, Hague is a typical example of panic in the face of responsibility, a characteristic often to be observed in men otherwise rational and balanced.

Bourn' s Interests Turn to Banking

In 1888, with the idea of retiring from mining, Bourn sold sufficient stock in the Empire Mine to James D. Hague and his eastern associates to give Hague controlling interest. Although he retained his seat on the board of directors, Bourn had now divested himself of control of both the Empire and the North Star and was able to retire from active management of the mines. At that time, Bourn was beginning to pursue a number of interests widely removed from mining. In fact, the previous year, on April 17, 1887, the announcement below appeared in the Grass Valley *Union.* Apparently, Bourn was venturing into one of his father's early fields, the world of banking:

> **The First National Bank**
>
> "The First National Bank of Grass Valley," under charter from the U. S. Treasury Department, will open for business Tuesday in the Holbrooke Hotel block. The establishment has been neatly fitted up, the desks, tables, counters and chairs being of solid oak, varnished, and while plain in appearance are solid and durable. In addition, there is a large safe of the Hall pattern, with combination lock, and a vault protected by a massive iron door, also fitted with combination locks. The bank starts with a paid up capital of $50,000, and is entitled to issue that amount of notes as a circulating medium, which

> are secured by a deposit of Government bonds in the U. S. Treasury. The establishment of the bank with the amount of capital named, is an evidence of the increasing business of the town, and is an institution that will find an ample field for its operations, and a great convenience to the public requiring banking facilities. The president of the bank is David McKay, who has been prominently connected with business here for some time, as Superintendent of the Empire Mine, and in merchandising. Mr. H. D. Andrews is cashier, a gentleman, who has frequently been a visitor here, and is well known to many of the business community. The other Directors are W. B. Bourn, Jr., J. H. Hammond, and Chas. E. Clinch, all well known and prominently identified with the mining and business interests of Grass Valley. The bank will be an important adjunct to the business of the town, and will undoubtedly be profitable to the stockholders. Its advertisement will be found in today's *Union.*

About that same time, Bourn's attention was being directed more and more frequently to certain disturbing events in Napa Valley.

The free and easy life style and general affluence of Northern California citizens following the gold rush days had stimulated a growing demand for fine wines and champagnes, and there was a rapid increase in acreage devoted to wine grapes. By 1880, hundreds of vineyards dotted the coastal valleys, with the Napa area, home of the estate vineyards of Madroño, a favored location.

Quite a few small wineries were in operation throughout the valley, and many growers marketed their crops under their own labels through these wineries. A majority of growers, especially the larger ones, preferred selling their wine on the bulk market in San Francisco. Almost none of the local wineries had the facilities for storage of large quantities of wine, and even with such facilities, the growers needed to convert their annual crops into cash as quickly as possible. As the number of vintners increased, so also did the supply of wine, and the bulk market became extremely competitive.

The wine merchants quickly took advantage of this situation. There was rampant collusion and price fixing, which forced prices down. To add to the distress of the growers, until the late 1880s the banks would not lend money on wine in storage, so vintners were forced to sell their wines very young at outrageously low prices. By 1887, Napa growers, demoralized by these dictatorial practices, and with revenue losses approaching $50,000 annually, banded together to seek relief from a steadily worsening situation.

Realizing that the power of the unscrupulous San Francisco combines must be broken, W. B. Bourn proposed construction of a gigantic wine cellar equipped with the finest cooperage. With a facility that size, the growers could store their wines at nominal cost, bring them to market at full maturity, and, thus, have a premium product to market. As banks began to accept wine in storage as collateral, growers also could borrow to finance future operations.

To finance a project of this magnitude required a fully cooperative investment program. The growers pledged 5 percent of their crops for three years, and together with capital provided by Bourn and Everett Wise, a young business friend of Bourn's, the resources were raised to proceed with the project.

Time was all important and Bourn personally ramrodded completion of the building project in less than two years, creating the imposing edifice known as "Greystone Cellars." It still stands, just north of St. Helena, as a monument to the courage of the Napa Valley growers and to the drive and determination of W. B. Bourn.

The project was successful from the start, but unfortunately, within a few years the vines of Napa Valley were so decimated by phylloxera, a vine-killing insect, that no surplus wines existed to cause a marketing problem. By 1894, Greystone had become a white elephant. There was a series of owners over the next 50 years, but none was able to utilize the cellar's huge capacity successfully.

In 1945, the Christian Brothers leased space in the cellars, and by 1950 had nearly a million gallons of wine in storage. That year, the Christian Brothers purchased Greystone, extensively restored it, added facilities for production of champagnes and other sparkling wines, and operated it successfully.[6] On August 10, 1978, the winery was placed on the National Register of Historic Places.

Mother and son were both engaged in building projects in St. Helena in 1888. Almost simultaneously with the start of the construction of Greystone Cellars, Sarah's beloved Madroño, located three miles away, burned to the ground. Salvaging a few carved oak doors and some architectural stonework that had escaped destruction, Sarah began rebuilding a few years later.

This was the beginning of many building years for W. B. Bourn, Jr. In 1890, less than a year after completing Greystone Cellars, Bourn commissioned Willis Polk, a friend and well-known San Francisco architect, to build a palatial town home at 2550 Webster Street in San Francisco. The house was not completed until 1895-96, however.

Also in 1890, Bourn became a director of the San Francisco Gas Company, holding that post for many years and eventually succeeding to its presidency. An 1890 directory also lists him as president of the American Powder Packing Company, with headquarters at 401 California Street.[7]

Engrossed in these activities, he had not yet become homesick for the smell of blasting powder, the clatter of stamp mills, and the general challenges of mining operations, but the time was drawing near when he would once again become active in Grass Valley mining.

NOTES AND REFERENCES

1. *Bonanza—a* rich body of ore, prosperity;
   Borrasca—very low grade ore deposits, not possible to mine except at a loss or break-even.
   Both words are borrowed from Mexican miners. When a mine is not in pay ore, or the vein has pinched out or disappeared, it is in "en borra" or "emborrescada" or "borrasca." As one hears it on the Pacific coast, it implies ill luck or hard times, coupled with a stem resolution to keep pegging away. The antithesis "bonanza"—a large body of

pay ore—has come to mean especial prosperity. There is a cheerful proverb of the Mexican silver miners that runs: "As many days as you spend in borrasca you will surely spend in bonanza." From *The Story of the Mine,* by Charles Howard Sinn; D. Appleton and Company, New York 1898.

2. Conway, Marian F., *A History of the North Star Mine,* Grass Valley, California, 1981. Reprinted by the Nevada County Historical Society, Nevada City, California, 1989.
3. *Ibid.*
4. John Hayes Hammond (1855-1936). At that time (1884), Hammond was one of the country's most promising young mining engineers. He later became world famous for his development of African gold fields in the Transvaal and Rhodesia (now Zimbabwe). He was a nephew of Jack Hayes, famous in the history of the southwest as the intrepid leader of the Texas Rangers.
5. Hammond, J. H. *Autobiography of John Hayes Hammond,* originally published by Farrar and Rinehart, New York, 1935. Latest publisher, Ayer Company, 99 Main Street, Salem, New York, 03079.
6. *The Christian Brothers Wine Aging Cellars, formerly called "Greystone,"* a publication by the Christian Brothers, January 1979. Heublein Company of Canada purchased the Christian Brothers winery in 1991 and sold the Greystone Cellars building to the Culinary Institute of America, a gourmet cooking school from Hyde Park, New York, in May 1992.
7. Spangler, Ray. *Under the Court House Dome.* The Country Almanac (San Mateo County) August 1975.

~ 6 ~

Cousin George

'Somewhat of a Mining Man'

When Bourn sold a controlling interest in the Empire, the new owner—James D. Hague—named George Starr as mine superintendent. For Starr, however, this was simply maintaining his current status, as he had been running the Empire for his cousin, Will Bourn, for some two years prior to the change of ownership. Starr's rise from the bottom rung of the ladder to the position of superintendent at the age of 24 deserves explanation, nonetheless.

To say that Starr's family relationship played no part in his rapid rise to the top would be unrealistic, but to imply that he did not merit his success obviously would be untrue, also. It was not simply a demonstration of his adaptability, innate intelligence, and capacity to handle men. Nor was it just an indication that the New England bloodlines of the Chase family exhibited the same drive and determination that marked the Yankee Bourns. All these characteristics were abundantly evident, but there were other elements that ran deep inside George W. Starr, deeper than heritage; factors that related, justifiably or not, to George's personal assessment of his place in the family.

The fact that he and his mother had been abandoned, and that he lost his mother when he was 12 years old, may have combined to leave lingering scars. Further, all his life George displayed great affection and respect for his aunt Sarah, but the fact that he was dependent upon her generosity throughout his formative years, and the realization that he was a poor relation in the middle of what became a very wealthy family, may have aroused a sensitivity in him that could not be assuaged, no matter how much genuine affection Sarah showered on him.

Whatever the underlying influences that molded his character may have been, they culminated in a deep motivation for achievement at the highest levels of excellence. George grew into young manhood with a compelling drive to excel at every task he faced. Degrees of importance simply did not exist for George. In his days as a mucker, it was as important for him to fill his ore carts as fast as men of greater physical strength as it was, in later life, to present—with irrefutable logic and convincing persuasion—his arguments for improved mining equipment to a sometimes hostile board of directors.

Starr took on the most difficult and dirtiest jobs without being prodded by the mine foreman. Never using his relationship to Bourn to gain an unfair advantage, he tackled every assignment cheerfully. Grasping the fine points of mining swiftly, and listening carefully to everything the old Cornish experts had to say, he made friends and gained rapid acceptance by the men. With an instinct for mining, and a flair for leadership and responsibility that could not go unrecognized, he quickly worked his way up to a shift supervisor's job.

Letters to Aunt Sarah

Excerpts from some of George Starr's correspondence yield, perhaps, the sharpest images of both hard-rock Empire and of a young man coming of age in the mining at the Victorian West. They also shed some light on a few of the complexities of a very complex gentleman. These letters were all hand written and the punctuation has not been altered:

To Aunt Sara—80 East 56th St., New York, N. Y., January 7, 1884. (Aunt Sarah was in New York with her daughters, who were attending finishing school.)

Dear Aunt Sarah:

I wish you a happy new year. I received your kind letter quite awhile ago, but I was laid up with a sore foot at the time and the place where I was staying had no paper or ink. Since Ihave started in to work I have had no time to do hardly anything, there being several hundred tons of rock on the dump when we started the mill, therefore, keeping it going night and day and Sundays and Christmas. The mill was started on the 7th of Dec. 1883. It is a beautiful mill, everything runs well, crushing much more rock at a less expense with better results than the old mill. The mill consists of one rock breaker, a machine that breaks all the rock to the size of one's fist. Then, dropping into a large ore bin it passes through shakers into self-feeding bins, doing away with 3 men, from there into the batteries, then over a long line of silver-electrified plates, then into six Triumph concentrators, then out of the mill, passing away as tailings.

It is now quite a pleasure to work in it—everything can be kept so clean. The (illegible) has charge of one shift and I have the other. This week I am night shift and it is not at all desirable for we start in at 5 P.M. and quit at 7 AM. the following morning. Then by the time one has had a sleep and meals it is time to go to work again but when day shift comes we can appreciate it.

I got quite a mash when I hurt my foot. A stem that I was guiding, owing to the improper tackling used, fell and rolled onto my instep bruising it fearfully, but, luckily, no bones were broken. I was laid up 3 weeks and now I have to apply liniment twice a day to it, for exercising it causes it to swell.

I will have to close now hoping all are well. I remain your nephew.

G. W. Starr

(Starr was apparently installing a heavy steel stem of one of the stamps in the new mill he was erecting in December 1883.)

To Aunt Sarah from Grass Valley, January 20, 1886 (on a letterhead of the Original Empire Mill and Mining Company).

Dear Aunt Sarah:

I received yom letter yesterday and was sorry to see my neglect in writing.... I hope you will forgive me this time and will do better hereafter.

In regards to the question of what society I mingle with and whether by force of circumstance or not; I, to a certain extent by force circumstances, but mostly by my own free will, associate with one young lady here, by name Miss Libby Crocker, continually. For by her good character and winning ways I have become engaged to be married to her. Will, being with you now, he will explain all to you, if not write and I will. She was born back East and brought up here, has always gone with the best of people here, ladylike in her manner, kind disposition and as good a girl as ever lived.

Believe me Aunt Sarah what I write is true. I am not a haram-scarum that thinks not what I do, but have weighed everything, have made up my mind and it is settled. By your wish will wait two years unless something happens. It would be better to get married sooner, by that time I will be 25 years old and if alls well will then be able to support a wife. I could do so now fully as well, as a number of my friends do, which is very comfortably, but I will wait two years.

We were engaged last August. I should have told you when I was in St. Helena, but somehow it would not come out, but if you will forgive me this time I promise that what happens hereafter you will be well posted.

The mines are all looking well. That paper you sent is true as far as the richness of the mines is considered to be. I sent the card to you containing an engineering view of Yosemite thinking it the most appropriate.

Love all from
George

To Aunt Sarah, Grass Valley, February 14, 1886

Dear Aunt Sarah:

I received your kind letter and was quite anxious to hear from you and to know what you would think of my doings of late. I am very glad that you feel I am doing what is not wrong and look at it in the light you do. It also has removed an anxious feeling from Miss Crocker. I will now work in good earnest and prepare for the reception of her and be in circumstances to make her a good home and live happily. She has already been an advisor and helpmate to me … she is so sensible and full of character one cannot help liking her if he tried.

Willie was up here last week with several gentlemen including Mr. and Mrs. Fargo, Castle, Hammond, and Sherwood. They spent a very pleasant week and left last Sunday morning. He has made all preparations for getting in water power and the work is going on rapidly. In about four months the Empire, North Star and several other smaller claims will be run by water which will ensure the running of these mines for years to come on account of the cheapness of the power.

The Empire is not yielding as well as she has done, but as soon as certain developments are made she will come out again—for the ore is there and when we can work it at better advantage it will be taken out.

The North Star is booming. She has a ten stamp mill running steady on splendid rock and is opening up fine in the lower levels as well the upper levels. I do all their retorting and melting and two weeks ago I melted for them a $9200 bar and yesterday a $4900 bar, besides she is producing lots of sulphurets[1] assaying $90.00 per ton.

I have tried to buy stock here among small stockholders such as 50 and 100 shares to the man, and they wont sell for $1.50. There are some buyers at $2.00. It will no doubt run up high this summer when she will pay dividends. All the other mines are doing well especially the Crown Point[2] which is taking our specimens continually.

All are well here. Love to all from George.

To Aunt Sarah at the Dakota Arms, New York City, March 30, 1886

Dear Aunt Sarah:

Since I wrote you last I have been on the go as we are making a great many changes here and improvements. David

McKay[3] is at the North Star almost all the time, keeps me quite busy … The mines are all doing well and a great deal of work is being done throughout the whole district. The largest scheme on foot now is putting in water power at the Empire and North Star. The pipes are all distributed and connecting is going on at the upper end. It is a fine piece of work and in one month more everything here will be driven by water. The additional 20 stamps are being put up and before long a 40-stamp mill will be running at the Empire and 30 stamps at the North Star. Willie Bourn is well-known now as a wrestler.[4] He has proved that mining in Grass Valley is in its mere infancy, but is opening up several dead mines, getting in water power, and rumor says he is going to bring in electric lights. He is very busy.

David has a little boy which I have not seen yet, but, from what I hear, takes after his mother. Will has promised me the superintendency of the Empire as soon as the water is in. I am now acting superintendent, and when I am promoted I can easily accomplish the duties of superintendent as I have had charge of the mine for the last six months. All are well I wish you to remember me to all. I remain your affectionate nephew —George

While John Hayes Hammond was Bourn's consulting engineer, he and George Starr became close friends. Hammond recognized a good mining man when he saw one, and in Starr he saw both competency and reliability. He took George to Mexico in 1885 to help him examine some mines about 300 miles south of Pueblo. In his autobiography, Hammond describes an encounter he and George had with six bandits, one of the many bands of robbers that roamed throughout Mexico at that time.

After he and Starr had worked hard in the mines for 10 days collecting ore samples to carry back to the States for assay, they loaded a wagon with bags of their rock samples, along with baggage containing clothing. With a four-mule hitch, Starr, Hammond, and a mule driver started back north. They had hardly gone 15 miles when Starr noticed all the baggage that had been loaded in the rear of the wagon was gone and the straps and ropes used for tie-downs had been cut with a knife.

Loss of their clothing might have been tolerated, but the loss of their ore samples was serious; they would have to be retrieved or the men would face a delay of several weeks while they returned to the mines to dig out more rock. Backtracking through a light rain, they found footprints in the moist earth and surprised the bandits squatting on their heels in a field of sugar cane, tearing open the sacks. Using a sawed-off shotgun to emphasize their demands, they forced the six bandits to carry the sacks back to the wagon. The banditti were warned off with dire threats if they should show up again, and Hammond and Starr proceeded to Mexico City and back to Grass Valley without further adventure. This incident only served to draw the two men closer together.

Starr Goes to South Africa

In 1893, the Barnato brothers of London offered Hammond a previously unheard of salary to assume command of their mining operations along the famous Witwatersrand of the Transvaal, South Africa. Believing it was certain to become the largest mining camp in the world, Hammond took with him a staff of the best American mining men he could assemble, plus several brilliant young English geologists. George W. Starr was included in this elite company. They were the men who developed and managed the world's greatest gold fields; first for the Barnato brothers in the Transvaal, and later for Cecil B. Rhodes in Rhodesia (now Zimbabwe).

For five years, Starr managed a large mine on the outskirts of Johannesburg. One of his letters to William B. Bourn, Jr., in 1894, reveals the huge size of those mining operations. In the letter, George also drops just the slightest hint of a personal factor that may have nagged at him for many years; one that, perhaps, was the spur that drove him to set such exceedingly high standards of personal performance.

Letterhead of the Primrose Gold Mining Company, Germaston, near Johannesburg, July 15, 1894.

Dear Will—I received your interesting letter some time ago and would have answered it sooner but I have been so busy that I could not find the time to do so.

Well, at last Will I have got to be the general manager of a large mine. I might be slow, but I get there just same; and have proved to these Johnnies here that I am somewhat of a mining man myself, even if I cannot show diplomas from several large mining schools.

This mine is now the second largest profit-paying mine in the Rand. We average about $11,000 to $13,000 profit per month, and with the additional 60 stamps, making 160 in all, we will pay from $18,000 to $20,000 profit per month. I never worked so hard in my life and will have to work harder yet because I now have no consulting engineer. Hammond and Clement have left Barnato brothers and I come directly in contact with the board of directors.[5] How long I will last under their directing I cannot tell, but I will try to get on with them and, if not, well, Hammond will give me a good mine under him, so I will get along alright anyhow in this country.

I have 200 white men and 2400 Kaffirs[6] working, and the mine has a life of from 12 to 15 years at the rate of 160 stamps. About $250,000 will be spent this year for a new plant. We have been erecting a number of additions already, such as a 25 drill plant, new hoisting plant, and tailings intermediate settling vats for collecting tailings from the mill direct before treating with cyanide. A great many other things have to be built yet, such as the extra 60-stamp mill, a crawl tram line from the railroad, white men's quarters, the finishing up of the large work shops, the arranging of underground haulage through the main incline that goes off from the vertical shaft. There is a great deal to attend to and of course it makes a fellow wrestle to keep up his end.

I will send you by the mail the *Standard and Diggers News* showing the output for June which amounts to over 168000 ounces at about $17.00 per ounce.

This is a great mining camp and is becoming greater every day. I wish you could come down here for a trip and often wish that you could own some of these great mines. I am keeping an eye open to lease some mine here and if I can do so no doubt I will do well in it.

Last Fourth of July I raised a large American flag and after dinner set off some fireworks.

Give my love to Aunt Sarah and all the folks and trusting you will always be well and prosperous I am

Yours

George

Some 18 months after that letter, in the late afternoon of February 7, 1896, a very hot day in Transvaal, the shades were drawn in George Starr's bedroom. Around the bed, a group of figures moved in a cadence to the sound of a muffled drum from somewhere outside the windows.

Slowly turning, round and round, moving forward, moving back, bending low and stretching high, swaying gently, for more than half an hour, the dancers circled the bed, quietly, rhythmically, sensuously. In the subdued light, there were occasional flashes of gleaming white teeth and glistening black skin. Except for the modulated throb of the drum, the only sound was the soft scuffing of bare feet on the reed matting of the floor. Local tribesmen, in full native dress, were performing their "Dance of Happiness" in honor of Bwana George's new baby, as she lay sleeping soundly beside her mother.

Dorothy Starr may have had many festive birthdays in later years, but none so bizarre and dramatic as the one she slept through the day she was born.

There seems to be little doubt that George Starr's flair for winning the respect and affection of his men had not diminished when he traded the Irish, Italian, Comish, and Welsh mining crews of Grass Valley for the Bantu tribesmen of South Africa.

For some years, the political climate in South Africa slowly but steadily worsened for American and English owned investments. After the Jameson raid against the Boer republic, death sentences were issued to almost everyone even remotely involved. Successful mining operations became a virtual impossibility. It was time for George Starr to bring his family home.

The Starrs arrived back in Grass Valley in 1898 and George was promptly offered the post of mine manager at the Empire. Under Starr, the mine enjoyed its most profitable years, with his management ending in 1930, shortly after Bourn sold the mine to the Newmont Mining Company.

Starr stayed on at the Empire for several years as a consultant to the new Newmont management. He then retired and moved to San Francisco where he died June 21, 1940, at the age of 78. He was cremated at Cypress Lawn.

As a final sidelight on George Starr's character, his granddaughter indicates that, although he was a member of the Pacific Union Club of San Francisco, to her knowledge he only attended once. Except for some of the local Grass Valley community organizations, George was something of a loner. It is believed that is why he built a stone cabin on Osborne Hill, not far from the Empire mine. He did enjoy talking to the many men who came from all parts of the world to discuss mining with him, but left to himself, he preferred to read quietly in his own private hideaway.

NOTES AND REFERENCES

1. Sulphurets (modem spelling, "sulfurets"). Also frequently called "mineral concentrates." Primarily, they are Iron Sulphide (Fools Gold) and Lead Sulphide (Galena). The ores in the Grass Valley area ran about 2 percent to 3 percent sulphurets. In the 1880s, with gold at $19.75 per ounce, their value ran about $60 to $125 a ton.
2. The Crown Point was a small mine somewhere near the intersection of East Main Street with Idaho-Maryland road in Grass Valley. The mine was noted for producing museum class specimens of free gold.
3. David McKay married Mary Abigail Chase, one of Sarah Bourn's sisters.
4. Wrestler: vigorous pursuit of a desired goal. Persistence and refusal to accept defeat. Applications of effort to the point of exhaustion. These definitions, although somewhat different from those found in modem dictionaries, have a history going back several thousand years.

5. After getting all of the Transvaal mines well established under competent mine managers, Hammond left the Barnato brothers to work for Cecil B. Rhodes farther north in what eventually became Rhodesia, now Zimbabwe.
6. Kaffirs: members of a group of South African Bantu speaking people.

~ 7 ~

SARAH'S DAUGHTERS

MAYE BOURN

Less than a year after joining her husband in 1854, Sarah gave birth to her first child, Mary Champney Bourn (later called Maye). In her early years, Maye attended public school in San Francisco, but when their country home, Madroño, was acquired, Maye (then about 14) and her sisters were placed in the hands of special tutors and governesses.

With girls of widely different ages, Sarah was not disposed to see them leave home one at a time for their higher education. She wanted them to go as a group, and she intended to go with them. Maye, therefore, was in her late twenties when the time finally arrived that all four girls could enter an eastern finishing school. Sarah rented a large apartment for herself and her brood in The Dakota Arms[1] on Central Park West, just across from New York's Central Park.

By that time, Maye had become a part of San Francisco society, had developed a beautiful soprano voice, sophisticated conversational skills, and a flair for stories of all kinds.

At one point in Maye's adult life, she met, fell in love, and became engaged briefly to a Russian gentleman. They corresponded at length, but nothing came of the affair. Years later, in 1895, Maye married James Ellis Tucker of Winchester, Virginia.

The entire Russian incident seemed to have dropped from family memories until Maye died in 1947 at the age of 92. At that time, F. Bourn Hayne and his elder brother, Alston, were named co-executors of the estate and had to do a great amount of cleaning up and sorting through her possessions. While going through Maye's desk, Alston discovered a sizable packet of letters from her early lover, the Russian, neatly tied with a ribbon and bearing a note on top saying: "Burn these after my death." Alston read a few of the letters, and then without saying anything to his brother, Bourn, destroyed the entire packet. When Bourn learned of this, he was aggravated and unhappy because he felt they should have been saved.

(When Mrs. Hayne related this story to the author, she asked: "Now what is your moral obligation in a situation like that? Should a person's expressed wishes regarding disposition of private papers after their death be respected?" Most of the arguments, pro and con, were either decidedly emotional, or just plain romantic. Looking back, from the perspective of a number of years, perhaps the most convincing point was made by Mrs. Hayne herself when she observed that, "People used to write something meaningful in their letters during those Victorian years. The telephone was just in its infancy, so letters were all we had, and those letters were lengthy and filled with thoughtful observations, not only of a personal nature, but on life generally and the world around them. Perhaps they should be preserved for future history students." Mr. Hayne sat quietly throughout the telling of this story and the discussion that followed. Finally, when it appeared that the discussion was going in a circle, he cleared his throat and said: "It was history, and you don't throw history away.")

Despite her "lost love," Maye had a long and happy marriage to James Ellis Tucker, who was born in Winchester, Virginia, October 25, 1844. When the Civil War started, young Tucker was in a boarding school in Switzerland. For a short time, he followed the progress of the conflict through letters and newspapers, but with his southern passions aroused, Tucker ran away from the school and made his way to France to meet his father, who was there on a special diplomatic mission for the new Confederate states. At his father's direction, James went to London to receive secret papers for delivery to the Confederacy.

By that time, the Union Navy was beginning to halt and board all ships suspected of carrying supplies to the South, and Tucker's ship had some difficulty eluding the Yankee ships lying off the English coast. Once through the blockade, the ship set a course for the Carolinas, only to find that southern ports were closely guarded by Union warships. Changing course slightly, Tucker's vessel slipped into the port of Havana where they were bottled up by several Union gunboats. The naval ships had been patrolling the coast just outside the bay and one of them promptly took up a station close enough to spy on the Confederate bound ship. Since they were in neutral Spanish waters, there was no danger of attack, but Tucker's ship was at the mercy of the gunboats if they left the protection of Havana's harbor.

At that time, a great Spanish celebration was underway, and the City of Havana was holding a state ceremonial ball in honor of the occasion. The Mayor of Havana sent invitations to the officers of Tucker's ship, and also to the officers of the Union gunboat. Both ships courteously declined, of course, but the skipper of Tucker's vessel saw a glimmer of hope in the situation.

As twilight descended, the ship's captain had large iron plates brought out of the hold and covered a good part of the deck with them. Then, a keg of powder was poured in a heap on the plates. When it was touched off, there was a "whoosh," and sheets of flame leaped skyward. Observers on the Yankee gunboat saw and heard the explosion. It was immediately assumed that a serious explosion and fire had wrought great damage to their prey, and officers of the gunboat donned dress uniforms quickly and made for the fandango in Havana.

To add to the deception, a few of the officers from Tucker's vessel also showed up at the ball wearing solemn, worried faces. Stiff greetings were exchanged, and then both parties proceeded to join the drinking and dancing going on around them.

One by one, however, the officers of the Confederate ship quietly drifted away from the ball and made their way back to their vessel. Under cover of the dark night and a greatly reduced watch on board the gunboat, Tucker's ship slipped out of Havana harbor and set sail for a Florida port, which they reached without further incident. From Florida, Tucker made his way to Confederate headquarters and delivered his documents to Jefferson Davis. He then enlisted and fought for the South throughout the remainder of the war.

As soon as the conflict ended, Tucker went to Mexico, where he served briefly as a courtier at the court of the Austrian, Duke Maximilian. Maximilian, a French puppet, had been placed on the throne of Mexico by Napoleon III, who saw the Civil War as an opportunity to make an "end run" around the Monroe Doctrine and occupy Mexico, still weakened by its war with the United States, in which it lost half of its territory.

With the Northern victory in the Civil War, Napoleon found his forces occupying Mexico in an untenable position and recalled them as speedily as possible. Maximilian and his generals were executed by a Mexican firing squad, and Tucker returned to the United States.

By then, President Lincoln had been assassinated, and Tucker's father, who had long been a target of vicious political foes and the yellow journalism of the day; was accused of involvement in the heinous crime. Sensing the temper of the times, Nathaniel Tucker fled to Canada, where he stayed for almost two years, until the United States government in Washington completely cleared him of all charges.

When James learned what had happened, he went to Canada, visited with his father, and then returned to the United States. Almost penniless, and the South in ruins, his thoughts turned toward a possible future in the rapidly expanding state of California.

For some months, he worked in Utah to raise funds sufficient to get him to the West Coast. Once in San Francisco, he obtained employment as a purser on a ship in the China trade that specialized in bringing Chinese coolies to San Francisco. From there, they were sent to railroad construction camps, where, along with Irish immigrants, they formed the major labor force in building a rail link to the East, a project then being carried forward at a furious pace.

After a year or two at sea, Tucker gave up his purser's job and left San Francisco for work in the mercury mines above Clear Lake, California. There he met and married his first wife, Laura Harris. They had two sons, William and Beverly. A few years after the birth of the last child, Laura died, leaving Tucker free to wander the hills and valleys of the surrounding area. He became a great friend of Lily Hitchcock Croy, a woman living in the northern Napa Valley, who was noted for her kindness and interest in southerners. Lily's diary contains a number of references to Jimmy Tucker's frequent visits. It must have been on one of those visits to the Napa Valley that Tucker met Maye Bourn. They were married in 1895. Maye was then 40 years of age, and Tucker was 51.

Maye was very fond of all her in-laws. The Tucker family was prominent in Virginia and particularly so in the Episcopal Church. Several family members were missionaries in Japan, traveling back and forth several times a year. When they arrived in San Francisco on those journeys, they were usually entertained at Sarah's home at 2030 Broadway, and invariably Maye entertained them lavishly with champagne and all the trimmings at the Town and Country Club and with dinner at the finest restaurants.

All her adult life, Maye Bourn Tucker was an extremely extravagant woman and spent money like water. She never opened her bills, but instead just threw them in a box under her bed. Eventually, of course, mother Sarah, or sister Ida would have to pay the bills.

Maye Bourn Tucker died in San Francisco on November 2, 1947, at the age of 92. Her husband, James Ellis Tucker, died in San Francisco, February 24, 1924, at the age of 80.

SARAH FRANCES BOURN (ZAIDEE)

The second daughter of Sarah and W. B. Bourn, Sr., was born in San Francisco, June 5, 1859, and was known as Zaidee all her life. Apparently, she never used the name Sarah, because her tombstone carries the name "Zaidee Frances Bourn Moody."

Little information on Zaidee is available. She married Horace Moody of New York, the brother of Agnes Moody Bourn, wife of W. B. Bourn, Jr. Zaidee had no children. She became ill at Madroño and was taken to San Francisco where she died October 15, 1898, at the age of 39.

IDA HOXIE BOURN

Sarah's third daughter was born in Brooklyn, New York, April 2, 1864. As noted previously, Sarah was making her first visit to family and friends since leaving the East in 1854.

Ida had a reputation as a "string saver." Fancy gift wrappings were much in vogue, and when beautifully wrapped packages arrived for Ida from the City of Paris, Shreves, the White House, and other fine shops, she carefully removed the wrappings and their content, then painstakingly rewrapped each box, endeavoring to retie the ribbons in their original folds. After Ida's death, F. Bourn Hayne discovered dozens of artistically wrapped boxes in closets and under beds. On the bottom of each package—in small, neatly printed lettering—were the words "Empty Box."

A lifelong, close relationship existed between Ida and her cousin, George Starr, and his wife, Elizabeth. George had a strong interest in automobiles and, in the early days, owned a Stanley Steamer. Ida was always ready to travel, and it was probably in the Stanley that she and George made a trip to San Diego, then a long, arduous trip.

George and Ida corresponded frequently over the years. As close understanding friends, their letters touched on both the happy and the tragic, as exhibited in the following letter:

> Grass Valley—January 1910
>
> Dear Ida— I arrived from the East, Januaiy 3rd, after a most delightful trip. Lib *(he called his wife "Lib" or "Libby")* and I often wish you were with us on some of our jaunts, for we know how you enjoy a little time off. We both thank you for your kindness in buying the present for Agnes and Will and would like you to send the bill to me or at least let us know the cost so I can make payment.
>
> I suppose you all saw the air show and it must have interested Aunt Sarah.
>
> Love to you all
>
> George

Grass Valley, March 8, 1910

My dear Ida— I heard from Mrs. Bray that Aunt Carrie was expected to live but a few days and have been expecting a telegram from someone to notify me in case of death. Having heard nothing, thought I would write you for information. Will you please write or wire me?

I have had several bad accidents in our mine during the past two weeks. Three men have been killed and several badly hurt. Today one man was killed by being blasted and two badly hurt.

Love to all the folks
George

Ida never married and died m San Francisco, December 23, 1951, at the age of 87.

MAUD ELOISE CHASE BOURN

Sarah's fourth daughter, Maud, was born in San Francisco, November 15, 1867.

The tragedies at 1105 Taylor Street—first the loss of her youngest son in 1872, and then the death of her husband two years later—turned what had been a happy home for Sarah into a house of sad memories. She could no longer abide the place and put it up for sale. For a number of years, she and the girls stayed in a hotel in San Francisco, but finally, she rented a house on the northwest comer of Gough Street and Pacific Avenue.

By that time, the girls were all adults and, with finishing schools and special jaunts to Europe all behind them, they found San Francisco increasingly attractive, particularly during the opera and theater seasons. When their social life began to dominate their interests, it was a signal that the time had arrived for a new, permanent home in the city. Thus, about 1892, Sarah purchased a large house at 2030 Broadway, and Madroño was used in the warm summer months—when city activities were at a low ebb—and other special occasions. Coinciding with the move to 2030 Broadway, a new chapter opened in the life of the family.

Maud Bourn became engaged to William Alston Hayne II of Santa Barbara. He was the son of Col. William Alston Hayne and his wife Margaretta Stiles from South Carolina, who had settled in Montecito, California, in 1867, with their six sons, where they were taught to recite classical poetry. His grandfather, a silver-tongued senator from South Carolina, was the adversary of Daniel Webster in their famous running debate for nine days on the floor of the Senate in 1830.

Maud's courtship and marriage to the descendant of this famous South Carolina family are best told by her son, F. Bourn Hayne, in his delightful work, *"En Un Tiempo."*[2]

> In 1891, a Festival and Battle of Flowers was proclaimed for the visit of President Harrison to Santa Barbara. The Festival proved such a booming success in every way that it received publicity in papers throughout the nation and was unquestionably the reason why my widowed grandmother, Mrs. William B. Bourn, I, packed up her two unmarried daughters, my mother Maud and her older sister, Ida, and took ship for Santa Barbara the following year, 1892. Taking my mother's two wheeled dogcart, tandem harness, and many trunks of clothes, the widowed mother arrived By-the-Sea with her two daughters to survey the matrimonial field.
>
> Ostensibly to discuss the cultivation of the olive tree and to order olive cuttings shipped north to her estate Madroño, … Mrs. Bourn I met William Alston Hayne IL He had also observed Maud Bourn driving her dogcart, tandem style, around the unpaved streets of Montecito and Santa Barbara. Father entered into the spirit of the Festival of Flowers with great zeal and in 1894, he and his brother Ben … acted as Grand Marshals for the Parade, all on flower-bedecked horses. The year 1894 also marked the beginning of a five-year engagement to Maud Bourn.…
>
> Gold fever swept the Golden State as reports came back from Alaska and the Yukon in 1898. Father had been building his adobe house in Montecito, "a regular old Spanish house, 70 feet square, plaza in the center, tile roof, etc." named "Las

Tejas" since 1894 for his prospective bride, Maud Bourn. The cost of building "Las Tejas" with its stone foundations had been so great that he finally sold the adobe to his elder brother, Robert Hayne, a successful lawyer and state supreme court judge. Freed from his debts for the building and with extra money in his pocket, Father fell for the exaggerated newspaper articles that told of the wealth in gold north of the Bering Straits, across Kotzebul Sound and Hotham Inlet and 300 miles up the Kobuk (or Kowak) River. The Hayne party was to travel the river in a 32 foot long steam launch Father had built in San Francisco which was capable of towing behind a string of rowboats filled with tons of provisions.

J. J. Hollister, Sr., (owner of the famous Rancho Santa Anita) also fell for the line of free gold north of the Arctic Circle and organized his own party of Santa Barbara miners to go north.

On May 8, 1898, William Alston Hayne II and his party left the dock in San Francisco aboard the sailing bark *Northern Light,* while Hollister followed aboard the sailing bark *Alaska,* with the steam launch, the *Agnes M Boyd.* Both parties built log cabins on the Pick River, a southern fork of the Kobuk River. Untold hardships followed for both parties. *"Why they left their homes in the South to roam 'round the Pole, God only knows"* (a la Robert Service). Father returned to marry my mother, Maud Bourn, in San Francisco on December 27, 1899, then returned north to work claims north of Nome in the Yukon Territory, but without success.

Two sons were born to this union—William Alston Bourn Hayne, November 26, 1900, in San Francisco, and Francis Bourn Hayne, September 6, 1903, also in San Francisco.

F. Bourn Hayne remarked to the writer many years later: "The whole adventure took over a year out of father's life and was an utter failure, but at least he brought back enough gold for two wedding rings."

NOTES AND REFERENCES

1. The Dakota Arms Apartment was a lovely, old-fashioned apartment hotel, always popular with members of New York's theatrical world. In more recent times, it has earned the dubious distinction of being the scene of John Lennon's (the Beatles) assassination.

2. Hayne, Francis Bourn. *En Un Tiempo,* Early Days of the Society of Los Alamos. Schleck Printing, Inc., Napa, California, 1979. This privately published book was written to preserve for all time the history of the Society of Los Alamos of Santa Barbara County, and to honor in picture, poetry, and prose the gallant spirits of members of the Society. The Society of Los Alamos was formally organized in 1910, but the roots of the organization go far back to the early days of a school for boys established by the Brothers of the Mission Santa Barbara. As F. Bourn Hayne states in *En Un Tiempo:*

   > It is my strong conviction that the Mission School, the Colegio Franciscano, with its mixture of old Spanish blood and new stock formed the grass roots of the Society of Los Alamos. Many of the names of the charter members of the Society, and the early replacements, are listed in (the) roster of students attending the Colegio. The boys, in turn, learned to speak and write better Spanish and English, and more importantly, to like each other. A boyhood bond was formed that lasted throughout their lives. Although the Colegio was closed in 1877, bonds of marriage, ranching, roundups, and fiestas kept the original bunch together in the 1880s and 1890s. These were the "good old days" which Society members would recall in their stories of *En Un Tiempo,* "Once Upon a Time."

~ 8 ~

The Family Historian

The costly and fruitless prospecting adventure in the Alaskan gold rush of 1898 left William Alston Hayne in very reduced circumstances. His large groves of olive trees in the Santa Ines Valley were not providing a viable income, and for several years he eked out a living in the province of Durango in old Mexico. Suddenly, he received a letter from his wealthy brother-in-law, William B. Bourn, Jr., who had just purchased a ranch about 12 miles west of San Luis Obispo.

As F. Bourn Hayne remarked, "When father received Uncle Billie's letter, offering to fix him up with a ranch, he jumped at the chance, and that was how father found himself in a damned sandy trap of lupine, scrub oak, and jack rabbits. Father worked the ranch until 1915."

Sarah's home at 2030 Broadway was very spacious—large enough to accommodate all her daughters, married or single. Maud and her two sons lived there, even after her husband was settled on the ranch. She visited the ranch for only a few months each summer. She much preferred San Francisco to the ranch, where the sand was so deep that automobiles were useless, and the only means of traveling the 12 miles between the ranch and San Luis Obispo was by buggy or saddle horses. Normal amenities were practically nonexistent, and there were no schools available for the boys.

They were a big happy family for many years, until the earthquake of 1906. The Broadway home had no damage, but the family, including the Hayne family, packed up and moved to Madroño until the trauma of the shock wore off.

F. Bourn Hayne had happy memories of his early childhood at both San Francisco and the ranch. His grandmother's house

(2030 Broadway) was right next door to the house of the very famous Mrs. Eleanor Martin, a prominent and influential member of San Francisco's elite families—a grand dame of society, indeed. Hayne remembered a birthday party at her house, along with many other family anecdotes, in his discussions with the writer, as related below:

I THOUGHT IT WAS A PENNY

Although she lived only about 25 yards from grandmother's house, grandmother never went to her home and Mrs. Martin never visited grandmother's house. Mrs. Martin had a grandson, Charlie Martin, who was about my age, and twice they asked me to see motion pictures with them. We drove down in an open barouche to the Savoy theater to see a picture about penguins. This was the time Amundsen reached the South Pole and they were releasing stories and pictures of his discoveries. Charlie's mother went with us. She wore a large brimmed picture hat and looked very fine as we pulled up in front of the theater and were helped out by the driver.

At one of Charlie's birthday parties (the fifth I think) they invited me to his party. They had a big cake and there were lots of nurses around to watch the small fry. I got a five-dollar gold piece in my cake. It thought it was a penny, burst out crying, and threw it in the corner. I had never seen a five-dollar gold piece.

LIFE ON THE RANCH

When I was six or seven years old, I had a white saddle horse that my father rented for the summer months for me. Every Saturday, it was quite an excursion to drive in the surrey 12 miles to San Luis Obispo to buy supplies, then we had to drive the 12 miles back again. For the first four miles from the ranch, you drove in deep sand and then a rough road the rest of the way. It was slow going. We had to buy everything in quantities, like 100-pound sacks of sugar and 50-pound sacks of flour so we would have enough groceries. We had to have food for the hired hands as well as ourselves. There was no electricity, no gas. There was a windmill, and a passable well up to a five- or six-hundred gallon storage tank. The windmill was the only bit of machinery on the place.

Because it was such a long trip into San Luis Obispo for supplies, we tried to live off the land as much as we could. Father supplied the Mexican hands with either shotguns or 22 rifles and they would ride out into the brush with my brother and me. We would run our horses through the brush, stirring up a lot of dust and chasing out the jack rabbits, and—wingo!—the Mexicans would shoot them. Then they tied the dead rabbits to my saddle, because I had a stock saddle with a lot of strings attached, and before you would know it, I would have 8 or 10 bleeding rabbits hanging from my saddle. When I got back to the ranch, my horse's legs were crusted with blood and I would have to use our precious water to hose him off.

I learned to kill and skin jack rabbits down there. It was good experience. Now I wouldn't shoot a deer or anything, but when I was a kid it was necessary. It was the happiest time of my life in that damned black sand.

JOINING THE BOY SCOUTS

During my early teens, I continued to live with mother at grandmother's house on Broadway, with summer at the ranch and an occasional visit to Madroño.

When father finally gave up the ranch, he brought a very fine horse, Blackie, up to Madroño. Blackie was a horse of all work. He was a superb saddle horse, or he could be used to pull a buggy or work as part of a team of two or four. He could do just about anything that was asked of him.

When I was about 15 or 16, I joined the Boy Scouts and worked for a merit badge in Pioneering. For this project, I settled on the construction of a log cabin, about 10 by 10 feet. I cut pine trees for the logs I needed, and used Blackie to drag them to the building site. After building up the walls as far as I wanted with the logs, I built up the rest of the walls with old boards. For the rafters (I had no idea of stress and strain in those days) there was a lot of beautiful redwood molding that I found in the attic at Madroño, so I swiped those. The sill I made of old boards, and the roof was boards and tar paper.

It is noteworthy that F. Bourn Hayne had the distinction of being the first Boy Scout in San Francisco to achieve the award of Eagle Scout.

COLLEGE YEARS

In 1933, after I graduated from Curtis Cate's School for Boys in Carpenteria, I entered Harvard and mother went East with me. At one time, the famous mining engineer, John Hayes Hammond, had done a great amount of work at the Empire and North Star mines and the Hammond family and the Bourn family were socially quite close in San Francisco. Then, as usual, Uncle Will Bourn got into some sort of an argument with the famous engineer. Whatever it was about, I don't know, maybe Uncle Will objected to John Hayes Hammond's fee, but relations had been quite strained for several years. After I got settled in at Harvard, and mother in a rented room in Cambridge, she said it was absolutely ridiculous for us to keep up this family feud. She called the Hammonds, and John Hammond and his wife, Natalie, promptly invited mother and me up to their magnificent Tudor Gothic summer home near Gloucester. They had us several times and I thought John Hayes Hammond was just one swell man. He made me, a little freshman, feel like I was really somebody.

Come Christmas vacation of my freshman year, the Hammonds invited mother and me down to their palatial home in Washington, D. C., on Colorado Drive.

Hammond was wise enough not to get personally involved in politics, but he was much in demand as an advisor to the President, and to many influential Washington politicians. Just for mother and me, he invited ex-President and Chief Justice Taft and his wife to dinner. Taft was a wonderful fellow and she was a nice lady. Vice President Coolidge and Mrs. Coolidge were also there.

Hammond was aware that my mother, in her late teens, had a girlhood crush on General Pershing (in fact, I have a picture of mother and Pershing at West Point in 1884 when he was a cadet) so Hammond also invited General Pershing, but he couldn't come and it just about broke mother's heart. Hammond's house is now the French Embassy, so it was no shack.

After getting his degree in journalism from Harvard, F. Bourn Hayne wandered around Wyoming and Montana working as a wrangler at various cattle ranches. He always wanted to be a cowboy, and came close to realizing that dream. Then he took a job as assistant editor on a Nogales, Arizona, newspaper.

Life in that small, wild, western town, with its proximity to the border and to the continuing clashes between Mexican troops and the eternal insurrectionists roaming the mountains of old Mexico, suited Hayne just fine. He spent more than two years there, until—he related with a chuckle and a satisfied smile:

"I got a letter from John Hayes Hammond telling me it was high time I stopped fooling around and settled down."

In Nogales, he became a close friend of a soldier of fortune named Jim Hathaway. Jim had lived on the cutting edge for a long time and liked it that way. Years previously, just for the fun of it, he had joined up with Pancho Villa's forces as a volunteer machine gunner, and, as time passed, had formed strong bonds with many leaders in the Mexican gorilla movement.

While Hayne was on the Nogales paper, the rebel forces capitulated and were ordered to lay down their arms. Many of their leaders (all prominent Mexican citizens) would trust no one but their good friend Jim Hathaway. Rather than disarm within their own country, they crossed the border and surrendered their weapons to Hathaway.

Shortly afterward, when Hayne made a social visit to his friend's home, Jim threw back a blanket to display guns of all sorts piled high on his wife's sewing machine. Hayne finished the story this way:

"Come and take your pick, Jerky," Jim said. He always called me "Jerky." Well, there were ivory handled pistols and silver mounted pistols of all sorts in the heap. I didn't want any of those silver-mounted things, so I chose a little Smith and Weston .38. I still have it down stairs.

When I was on the paper, I was given the honorary title of deputy sheriff. That was in December 1927, and it was at that time that I received John Hayes Hammond's letter, which I still have in my files.

MATURITY FINALLY ARRIVES

After that I taught at Curtis Cates School, my old school in Carpenteria for two years. Then I worked in the architectural office of Arthur Brown and Company, the well-known architect, and I found that I liked this work very much. The first year it was on projects for the government. Hoover had scattered big government subsidized jobs throughout the country. Arthur Brown got the City Hall and the S. F. Opera House projects. He also received commissions for the Labor Building and the Interstate Commerce Commission building with a big connecting wing in Washington, D.C.

As an architectural trainee, I got $100 a month that first year, then in the second year, when government funding ran out, I didn't get paid anything.

In 1931, I got married. It was at the very depth of the depression, so I thought—instead of a futile search for a job that did not exist, I could afford to go back to architectural school in the East. We went to Boston and lived with my wife's parents and back to Harvard I went, to their architectural school, where I spent three years and two summer vacations. Harvard gave me credit for a lot of courses I had taken previously, because in those days architecture was a four-year graduate course with significant undergraduate requirements.

In 1915, F. Bourn Hayne's father left the ranch outside of San Luis Obispo, which he had been running for his brother-in-law, W. B. Bourn, Jr. He then sold off some valuable land in

Montecito, which had suddenly become a popular place for the construction of expensive homes. With the proceeds from that sale, he bought a ranch about 12 miles north of Marysville on the road to Oroville. He and his eldest son, Alston, worked the ranch, raising hogs and crops of hay and beans. They also had about 10 acres in olive trees, because *"Father was always crazy about olives,"* Hayne said.

While F. Bourn Hayne was in the East getting his degree in architecture, W. B. Bourn, Jr., crippled by strokes for many years, called Alston Hayne down from the Marysville ranch to assume responsibilities involving the sale of property around Crystal Springs Water Company and to act as manager of several buildings in the San Francisco business district.

When F. Bourn Hayne returned from Harvard, the ranch in Marysville was sold, and Hayne settled in Kentfield to establish his practice as an architect. He and his first wife had three children—a son, who died, and two daughters. One daughter, Mrs. Martha Hayne Talbot, lives in Switzerland where she and her husband have done outstanding work in ecology and in conservation of natural resources. The second daughter, Mrs. Sarah Simpson and her husband, own and operate extensive peach orchards in Northern California.

The writer first met F. Bourn Hayne in 1981 in St. Helena, where he lived with his charming second wife, Mrs. Ellen Hamilton Hayne. I enjoyed a most stimulating correspondence with Mr. Hayne, and his letters are a treasured part of my files. I was deeply saddened by his death in 1988.

~ 9 ~

BOURN AS A CAPITALIST

Born in 1857, just a few years after the days of the vigilantes, William Bowers Bourn, Jr., lived through an era that spanned the growth of San Francisco from a rude western village to a sophisticated metropolis of wealth, beauty, and refinement. The powerful magnates of finance, industry, and politics who triggered and molded that growth are central figures in San Francisco 's history. Bourn was a part of that era, and contributed to it in a significant manner. The four chapters that follow look at Bourn's life in the context of his times and environment.

Bourn's early involvement in the saga of the Empire Mine in Grass Valley was noted in Chapter 5. When he sold his controlling shares in the Empire in 1888 to his friend James D. Hague and completed construction the following year of the Greystone wine storage facility in St. Helena, Bourn freed himself from what had become two time-consuming enterprises. He could now pause, reflect, and chart a new course through the turmoil of San Francisco's financial climate.

In 1893, the brilliant mine manager, George Starr, was lured away to the gold fields of South Africa. The Empire was not paying well. It had encountered one of the barren zones that occurred several times in the mine's history. Bourn sat back and bided his time, and in 1896 he repurchased the mine from Hague. Two years later, when Starr returned to Grass Valley after five years in South Africa, he was promptly rehired by Bourn to manage the Empire. The barren zone was penetrated, reopening the vein. A modem ore processing facility was installed, and once again the Empire resumed paying handsome profits. From 1891 to 1928, more than $30 million in gold was produced.

The immense profit from the mine served as a cornerstone for Bourn's large personal wealth. He began investing heavily in public utilities, and in 1890 became president of the San Francisco Gas Company.

This was a fertile field in the days when San Francisco was rapidly becoming the leading city of the West Coast. Ferry boats, telephones, trolley lines, gas, electricity, and other public necessities were all run by a complex patchwork of competing companies. The moral business code of the day was cutthroat competition.

Electric lighting was introduced to California in 1879. As the arc lamp and the Edison incandescent gained favor, natural gas companies took up a defensive posture and consolidation soon became the name of the business game.

Thus, in 1896, the San Francisco Gas and Electric Company was formed under the presidency of the conservative Joseph B. Crockett. Bourn's San Francisco Gas Company was one of the important members of this consolidation. Crockett was challenged by two new independent companies founded by Claus Spreckels, the flamboyant sugar king, who was always looking for a good fight with his business competitors, or, better yet, members of his own family.

Spreckels was a self-made German immigrant. In search of greater sugar fields, he once traveled to Hawaii, where he met the easy going King Kalakua. The two men became poker playing buddies. One night Spreckels found himself the winner of a large part of Maui. According to family legend, his good luck came through the following ploy. He declared his poker hand had four kings. When he laid down the hand, he had only three. Kalakua asked where the fourth king was, and Spreckels replied: *"The fourth king, sir, is yourself!"* The King supposedly thought this was a wonderful joke and accepted his loss in good spirits.[1]

Early in 1899, Crockett and Spreckels chanced to meet at the staid Pacific Union Club during lunch time. Crockett was a super club man, past president of the Pacific Union Club, and member of the Burlingame and Bohemian clubs. The conversation supposedly went something like this:

Spreckels: *"Mr. Crockett, what are you going to do about the smoke from your power houses. The soot is blackening the walls of my building and filling my tenants' offices. It's a confounded nuisance."*

Crockett: *"Sir, I make it a rule to never discuss business at my club. If you will take the matter up with me at my office, I shall be most happy to consider it."*

The soot was never satisfactorily eliminated, and Spreckels never forgot the alleged slight, which spurred him to greater competitive heights.

In a typical Spreckels reversal, his son, Rudolph, managed to get himself elected to the Board of Directors of the San Francisco Gas and Electric Company (against whom his father was fighting). Rudolph immediately began a campaign of criticism, seriously questioning his own president, Crockett, who was ultimately forced to resign.

Into this melee stepped W. B. Bourn, Jr., who was then (1902) elected as the new president. In the ensuing battle, Bourn and the San Francisco Gas and Electric Company won out against the other competing companies. Bourn then took the next logical step. In 1903, he consolidated by buying out Claus Spreckels.

Bourn carefully reorganized his new company, but, alas, another strong competitor emerged in the form of the California Gas and Electric Company. More legal sparring and rate wars took place, until, on October 10, 1905, a final entity was created to be known thereafter as the Pacific Gas and Electric Company, still in operation today.[2] PG&E assumed control of both the California and the San Francisco Gas and Electric Companies. Bourn resigned as president and received a substantial amount of cash and bonds for his holdings.

Most publicity and biographical releases describe William Bourn, Jr., as "owner of the Empire Mine and President of the Spring Valley Water Company." Together, they comprised the primary source of Bourn's income and occupied the center of his business attention over most of his adult life.

William Bourn, Jr., as a young man

William Bourn, Sr.

Mr. and Mrs. William Bourn, Jr., on a stroll at Filoli.

Mrs. Agnes Bourn, wife of William Bourn, Jr.

A Bourn family portrait. *Standing, from left, Arthur Rose Vincent, husband of Maud Bourn Vincent; William B. Bourn, Jr.; Maye Bourn Tucker. Seated, from left, Maud Bourn Vincent, (standing in front) Elizabeth Rose Vincent, Agnes Moody Bourn, Sarah Esther Chase Bourn, Maud Eloise Chase Bourn Hayne, and Ida Hoxie Bourn.*

Photo courtesy of Filoli

Maud Bourn Vincent, daughter of Mr. and Mrs. William Bourn, Jr.

Arthur William Bourn Vincent, son of Arthur and Maud Vincent, and grandson of Mr. and Mrs. William Bourn, Jr.

George Starr, cousin of William Bourn, Jr., long-time superintendent of the Empire Mine, and "somewhat of a mining man" in his own right.

Libby Starr, wife of George Starr.

The Bourn "Cottage" at the Empire Mine in Grass Valley, California. Built in 1897, the Cottage served principally as a summer residence for the Bourns, and then for only short periods because of the noise from the mine operations.

Katy Moriarty was the housekeeper at the Cottage, and was said to be the real "boss" of the house, even though the Bourns owned it.

William Bourn, Sr., bought a country home in St. Helena, California, for his wife, Sarah, in 1867. Sarah built this palatial home known as Madroño, which burned down in 1888.

Photo courtesy of Filoli.

After the first Madroño burned, Sarah Bourn made plans to rebuild. The second Madroño was completed in 1903-1904.

Photo courtesy of Filoli.

William Bourn, Jr., commissioned Willis Polk to build this townhouse at 2550 Webster Street in San Francisco. The house was completed in 1895-96, and served as the Bourn home until after the Great Earthquake of 1906.

Willis Polk, famed San Francisco architect, designed many of the Bourn houses, including the Cottage at the Empire Mine.

Muckross House in Killarney, Ireland, was given by the Bourns as a wedding present in 1910 to their daughter, Maud, and their son-in-law, Arthur Vincent, an Irish nobleman. In 1932, the Bourns and Senator Vincent presented Muckross to the Irish nation as a memorial to Maud Bourn Vincent, who died in 1929 of pneumonia.

A portion of the mineyard at the Empire Mine showing the mine manager's office, the refinery, and the rescue station. The buildings were constructed from waste rock taken from the mine.

Another view of the Empire's mineyard, showing the headframe and stacks of logs to be used for timbering in the mine.

The Spring Valley Water Company also was to have a substantial impact on San Francisco. Bourn and his associates managed over the years to acquire virtual monopolistic control over the one item vital to the growth of any California city—its clean water supply.

From 1852 on, SVWC—which supplied water to the City of San Francisco—began purchasing land parcels around the Crystal Springs Lake area on the peninsula just south of the city. Its grip on the city water supply was not relaxed until Bourn finally sold it to San Francisco in 1930. He had been principal owner and president of the company since 1908.

'Bourn Luck' Backed by Research

Bourn never made a substantial investment without first acquiring a good knowledge of the problems, liabilities, and profit potential associated with the proposed investment. He sought the counsel of his successful business associates and engaged in long-range planning as standard practice. Those associates who coined the term "Bourn Luck" must have been unaware of the strenuous effort and extent of research and study that preceded Bourn' s business ventures.

An excellent example is a record of his 1903 trip through the Sierra Nevada to evaluate potential water sources. The story is told by William Bowers Bourn Ingalls, the "Willie" Ingalls of Brooklyn (see Chapter 3), who had turned to his wealthy cousin some years previously for a job. Thus, at age 50, he became Bourn's personal secretary. Always a faithful correspondent, he wrote to his brother Abbott, as follows (spelling and punctuation are shown as written):

> Dear Abbott:
>
> We have just returned from a most interesting trip covering about 10 days, and in that time we went over a great deal of ground.
>
> My last letter to you was from St. Helena written a few days after reaching there, so that I will continue from that

point. I spent about a week with Aunt Sarah at Madrofio and enjoyed it very much, interesting myself about the place and trying to accustom myself to the surroundings.

August 14. I returned to San Francisco and in the afternoon went with Ida to a symphony concert, which I enjoyed very much. The orchestra played very well comparing favorably with the orchestras with us. One selection was the Overture from Tannhauser which was beautifully rendered. The following morning Will, Mr. William H. Crocker, Mr. Joseph D. Grant and myself started for Grass Valley, arriving here in time for dinner at about seven. The ride up the Sacramento valley was lovely, the fields on either side of the road planted in wheat. As we reached the upper part of the valley we ran into the fruit section, vineyards and fruit trees on all sides, everything so extensive. I do not wonder that Californians have large ideas.

On our way up, Will talked with me about Grass Valley, saying it was only a mining town and that I must not expect much in the way of accommodations, that the Empire buildings were broken down and in bad condition as it did not or would not pay to rebuild them, and the houses were overrun with BB. *(This was not explained.)* I accepted what he said, so that you can imagine my surprise when I reached this cottage *(the Bourn Cottage at the Empire Mine)* to find a beautiful modem building with all the conveniences and lovely grounds about the house all beautifully kept with the loveliest roses and plants comparing favorably with some of the places at Tuxedo.

The mine is within a stones throw of the house and is located on the outskirts of the town, a drive from the station of about 15 minutes. The buildings are in perfect condition and are kept so clean, as a matter of fact I have been told it is a model plant and people have come miles to go over it.

Sunday morning, the day after our arrival, after breakfast we went down into the mine and to me it was a novel experience. We dressed ourselves in miners costume and then went to the shaft. The car was on a narrow gauge track on quite an angle or incline. The car itself reminds me of a long sled or toboggan. Each one was given a candle and at a given signal to the man at the hoist, we began to descend to the bowels of the earth. Our first stop was at the 1100 foot level, here we saw some of the miners working. While on this level we saw a mule used for hauling the ore that had not seen daylight for four years. Imagine living that length of time under-ground. After seeing everything on this level we returned to the shaft and continued down to the 2700 level where we got out and went into another tunnel. Will made some examinations here, pointing out to us the good quartz so that we could readily see the difference between desirable and undesirable. We then returned to the shaft and in due time reached the opening, and it seemed good to see day-light after being in darkness for about an hour.

We then went about the plant, seeing the different machinery working each doing its part in producing the precious metal. After lunch or at two thirty, we took a carriage and drove into town and then took a trolly to Nevada City a few miles north of this place. George Starr who is Supt. of the mine was with us, so we numbered five. Starr is a fine fellow and has had quite an experience for he was for five years in So. Africa and was in Johannesburg at the time of the Jamison Raid and his experiences read like a novel. He was consulting engineer for Barney Barnato for five years at a salary of $35,000 and expenses.

At Nevada City we took two teams and commenced our driving trip, which lasted about a week. Our objective point for Sunday night was Gaston Ridge which we reached Sunday night about ten o'clock. This is nothing more than the location of a mine, the place consisting of a few houses and a tunnel running into the side of the mountain. In reaching this place we go through some very fine timber land, White, Sugar and Yellow pine beside many other varieties of trees. The pines are large forest trees, much larger than with us, handsome and stately the branches about 100 feet from the ground. The bark of the tree of the yellow pine is very handsome of a brown, reminding me of a large crocodile skin. The country is mountainous and rugged boulders standing out in bold relief. I never saw such scenery, not even my trip through the White Mtns. compare with it.

We took supper at a place called Washington City, situated in a valley. On reaching here we went over a mountain I cannot give the elevation, but the drop was 2000 feet in four miles, so you can get an idea of the country. We reached Gaston Ridge about ten o'clock at night and we were a perfect sight, dusty and tired. I was dubbed Dusty Bill, and it was certainly well applied. We all had baths and finally turned in. The next morning we made an early start for Bowman's Dam, reaching there in time for lunch. This dam is situated at the head of a canyon, mountains on each side, forming quite a reservoir, holding thousands of gallons. The general view with the boarding house reminding me of views I have seen of Switzerland. In the afternoon we drove to two more lakes further up in the country, and here we had out first view of snow on the mountains and it was beautiful. I started with the party to visit another lake on foot but much to my chagrin I gave out, owing to the high altitude 6200 feet. I returned down the mountain to the house we started from and waited there for their return. We then returned to Bowman's Dam and stayed overnight. To give you an idea of the scenery, Will said that in some respects he considered the country we were in more interesting than the Yosemite.

The next morning, Tuesday August 18, we started early over the mountains, our objective point being Emigrant Gap which we reached about 5 in the afternoon. The views were magnificent and roads in some places fearful, one place in particular known as the Devil's Gate, just before reaching the Gap was almost impassable. We walked through this section for about ½ a mile, letting the drivers get the carriages through the best way they could. Long before we reached the Gap we could see the snow sheds of the Southern Pacific R.R., winding around the mountains which we had been through two weeks before on our way to San Francisco.

In going through this country I have thought so much of Bret Harte's stories, and in future shall read them with so much more interest. We finally reached Emigrant Gap, and there left our carriages, taking a local train to Truckee the toughest and wickedest town in the country, vice of all kinds carried on openly. We waited here for the over-land mail train taking it as far as Reno, where we stayed overnight. We had good accommodations with a tub adjoining our rooms at the hotel, and had a good night's rest. The next morning, Wednesday, we took the train for Carson City arriving there in time for lunch at the swell hotel of the place. After lunch we continued our driving trip, taking a three seated mountain wagon drawn by four horses. The wagon reminded me of a Government outfit, covered with canvas. The drive down the Carson valley was charming, the valley itself is very fertile, twenty miles wide and thirty five miles long, our objective point being Woodford's, where we stayed overnight and the next morning started for the Blue Lakes in the Sierras. All our meals had been specially ordered so that at each place as we arrive they were expecting us. It got to be a regular joke with us that we were not having anything to eat, but as a matter of fact we were living like lords.

It was suggested that one of the party write an account of the trip under the following title: "A trip over the Sierras or how we Suffered."

After leaving Woodford's we wound our way up the canyon and shortly entered three successive valleys, named Hope, Faith, Charity, the latter being the largest and most beautiful, carrying out the Biblical idea from Faith Valley, in looking west we saw Kit Carson's pass in the distance, which two days later we practically went through. We continued winding our way over the mountains and before reaching Blue Lake we passed over the divide which means that at that point the water sheds separate one going to the Nevada side of the mountains and the other to the California side. The trip from the start has been for the purpose of inspecting the water sheds of California with the idea of developing later.

We reached Blue Lake in time for lunch and spent the following two nights here, sleeping in a Log Cabin built about 25 years ago. The divide I have just spoken of happens to be the highest point of our trip 8700 feet elevation. The air was lovely and clear and from this point we had a very extended view and the quantity of snow in patches added much to the beauty of the picture. At the extreme end of the Log Cabin was a fireplace larger than yours at Tuxedo and this was finally lighted, throwing out a good heat, the light in the cabin was lovely. At this place the Standard Electric Co. of S.F. is building a dam, employing 130 men of many nationalities, including Indians. The Chief of the Washoe Tribe is a boss for the company and controls the Indians. His name is Pete Mayo and is quite a character. He called upon Mr. Crocker who is a large stock holder in the company and spent the evening with us. It was very interesting to hear him talk, for he spoke good English and knows the country, always having lived there. The Indians never sleep in the house so that the camp fires at night added to the novelty. The following day we went down to Deer Valley, some distance below the dam, to look over the sight for a dam that the company may build some time in the future, returning to our headquarters for lunch. We spent the afternoon in looking over the dam under construction and went to another lake, Meadow Lake by name, which is in this system. This lake is in a basin, the mountains on two sides coming down to the waters edge, rough bold granite, it was magnificent.

The following morning we made an early start, our destination being Bear River where we were to stay overnight. The thermometer when I got up was 48 degrees and it felt chilly. Leaving Blue Lake we returned to the divide, going as far back as Faith Valley, where we branched off and went through the canyon leading to Kit Carson's Pass. It was a very rough road and country, and the early settlers must have had a tough time making their way through. We saw the stump of the tree where Kit Carson carved his name, that portion of the tree, however, had been removed to the Forestry Department, either at S. F., or Washington. I think the latter.

On reaching the Pass we got out and went to a point where we saw the marks of the ropes on the trees where a block and tackle had been fastened to pull up the emigrant wagons. Certainly those men had a strenuous life. We took lunch this day at Tragedy Springs, a quiet little nook in a valley where on June 26, 1848 three men were murdered by the Indians. The graves are encircled by large rough boulders and on a large tree near by, the names are given, stating the facts. Mr. Grant was greatly interested in this place and a collection was taken up among the party, the object being to have a handsome brass plaque made and fastened to one of the large boulders near the spot. We stayed this night at Bear River in a primitive house, one of the stations of the Electric Co, visiting in the afternoon a dam located a mile from the house. From what I have written you can readily see that Water is the power of the country and the great question.

Leaving Bear River the next morning, we started for Electra, where the plant of the Electric Co. is located, reaching there in time for a late dinner at night, and this was the end of our trip. It was very interesting to have seen the hard running waters of this system and follow it along mile by mile, gaining in force resulting in running these enormous engines located at Electra. Will Bourn saw the electrical plant when in N. Y., which supplies the city and he considers the one at Electra far ahead of it.

We took lunch on this our last day at Mill Creek, a little station in the wilds of the forest about twenty miles from Electra. After lunch our party took horses going over a mountain trail and finally reached Electra. The views have all through been magnificent, and we have wished for you many times. Our first view of Electra was from the mountains and looking down into the valley from an elevation of 1300 feet we had a beautiful birdseye view of the entire plant. Before going down the mountain we got off our horses and sent our luggage ahead on their backs. We started down on foot, down the side of the mountain. The plant is extensive and shows an investment of an immense amount of capital. The house where we stayed is a very fine building like a country club, with all the conveniences. We went through the power house that night after dinner, and the thing that impressed me was everything was so clean, the immense wheels and all run by water power.

The next morning we had an early breakfast and a drive of twenty miles to Ione where we took the train. George Starr and I left the party at Galt, the others going on to the City, and we returning to Grass Valley. If you look on the map you will find we made a complete circle.

Sincerely yours
Wm. Bowers Bourn Ingalls

Bourn Fights S.F. Water War

In the early twentieth century, San Francisco went through one of its more colorful eras. Names of many aggressive capitalists filled the rich lore of San Francisco millionaires: Crocker, de Young, Spreckels, Fleischhacker, Haas, Zellerbach, Giannini, and others. The City's palatial homes on Nob Hill were built by silver and railroad kings. Unfortunately, her politicians were subject to a considerable amount of graft and corruption—more like greedy puppets dancing to the strings of the local "robber barons."

Through this assemblage of unbridled characters, William Bourn jockeyed his Spring Valley Water Company and managed to keep up with the best of them. Francis Bourn Hayne, William Bourn' s nephew, recalls the following story indicative of his uncle's business character:

> Uncle Bill Bourn had a fight with Mike de Young, who owned the San Francisco Chronicle. That was when Uncle Bill was President of the Spring Valley Water Company. The man who read the meters, reported to his boss that the Chronicle hadn't paid its water bill. I guess it was for several months, and finally it got up to Uncle Will's office that the Chronicle hadn't paid its water bill. It must have been four

> or five months before they would take it up to the president. So, Uncle Will said he wasn't afraid of de Young and his publicity, "Shut off the water. If Mike deYoung doesn't pay his water bill, shut off the water." So they shut off the water to the Chronicle, and the water closets wouldn't work, and no water would come out, and Mike deYoung was furious because he thought no one would dare confront him because of the publicity. Of course he paid up right away. I have heard he used the power of his press to get by with things, but not with Uncle Will.

One of the more unsavory relations at that time was between Eugene E. Schmitz, elected mayor of San Francisco between 1901-1907, and his infamous crony, "Boss" Abe Reuf, who had managed Schmitz's three campaigns and was in complete control of City Hall. San Francisco continued to grapple with the problems of its limited water supply. Some even went so far as to partly blame the Spring Valley Water Company for the dry water hydrants during the great fire and earthquake of 1906.

Reuf and Schmitz concocted a plan whereby an organization called the Bay Cities Water Company would tap Lake Tahoe and sell the water rights to San Francisco for $1.5 million. Out of this sum, Reuf was to receive a modest "attorney's fee" of $1 million, to be divided among "interested parties." Other payoffs followed on other deals.

In a periodic surge of morality, San Francisco ushered in a brief era of reform led by James Phelan, the former mayor: Fremont Older, the Editor of the *Evening Bulletin;* and none other than Rudolph Spreckels, the wealthy sugar king. The result was the celebrated "graft trials" that occupied center stage in San Francisco for several years.

The tactics of the Bay Cities Water Company were exposed, and their revelation destroyed its chances of becoming the dominant water company in San Francisco. The graft trials themselves began to wear thin on the general population, which grew tired of reading about the countless courtroom battles. A number of convictions were reversed. The business community, which first welcomed the reform measures, later changed its attitude toward an atmosphere of hostility after attention turned toward the "bribe givers," who often turned out to be men of standing and influence in the commercial and social life of the city.

San Francisco had tried on many occasions to buy the Spring Valley Water Company. One of the earliest attempts was in 1875, when William Ralston, of Bank of California fame, tried to peddle it to the city to help save his crumbling fortune. The city considered the price too high. In August of that year, the doors of Ralston's bank were closed, whereupon he resigned as President, went for his customary swim in North Beach, and drowned—either by accident or design.

In 1900, the voters adopted a new charter committing the city to a policy of municipal ownership of public utilities. The consolidation of the "Muni" railroad was begun and took years to complete. The water plan was to prove an even tougher challenge. By 1910, the city was using 40 million gallons a day, taxing its supply and facilities to the limit.

The city engineers turned their sights toward the inviting headwaters of the Tuolumne River in the northern part of Yosemite National Park, in a pristine area called the Hetch Hetchy Valley. For 10 years, an epic battle raged over the proposal to dam the Hetch Hetchy, reaching all the way to the presidential desks of Roosevelt (Theodore), Taft, and Wilson. The proponents were led by Mayor Jim Phelan and various Federal officials.

The opposition was led by John Muir, the dedicated and celebrated naturalist, and the editors of many national newspapers and magazines. Muir, whose disciples formed the influential Sierra Club, extolled the beauty of the area by exclaiming "No holier temple has ever been consecrated by the heart of man."

In the end, the conservationists lost. When the Raker bill was signed into law in December 1913, it provided for the construction of the dam and prohibited the city from selling any water to any private corporation for the purpose of reselling it.

Above the fray sat the imperial figure of William Bourn in his offices at the Spring Valley Water Company, sifting the pros and cons of civic versus personal gain. Soon he lined up squarely with John Muir, not because of conservationist zeal, but rather that he correctly saw the threatening challenge to his water empire. He went so far as to proclaim rather preposterously that San Francisco didn't need any more water. Furthermore, he added the interesting argument that if the water came from the Hetch Hetchy it would be polluted by the rising number of tourists in Yosemite Park.

After losing out to the Raker Bill, Bourn continued to thwart the city's proposal to consolidate its water plan. Between 1910 and 1928, five bond issues designed to finance the purchase of the water company's property were submitted to the voters. In each case, the asking price was considered too high, and all were rejected. Not until 1930 did a sixth bond issue—for $41 million—receive the two-thirds majority necessary for the measure to carry.[3] As an interesting after note, the city found that after final approval it could find no purchasers for the bonds. A P. Giannini then stepped in with his Bank of Italy (later the Bank of America) and led a syndicate that purchased the entire Spring Valley bond issue.

Bourn had many other business interests, of course. One was his connection to Fireman's Fund Insurance Company, founded in San Francisco in 1863. A good share of his profits from the Empire Mine were invested in shares of the insurance company, which produced a substantial return. His father, William Bowers Bourn, Sr., had been president of the company in 1866.

NOTES AND REFERENCES

1. Stephen Birmingham, *California Rich,* (New York: Simon and Schuster, 1980), p. 61.
2. Charles M. Coleman, *Pacific Gas and Electric Company of California,* (New York: McGraw Hill, 1952), p. 85.
3. Oscar Lewis, *San Francisco: Mission to Metropolis,* (Berkeley: Howell-North, 1966), p. 23.

~ 10 ~

BOURN AS A PHILANTHROPIST

The intellectual climate toward philanthropy at the turn of the century was strongly influenced by the concept of Social Darwinism. Perhaps the man who best exemplified its ideas was Andrew Carnegie. In 1889, he published a provocative article in the *North American Review* that has come to be known as the "Gospel of Wealth." Carnegie had a problem to grapple with. In 1901, at age 66, he sold his steel business to J. P. Morgan. His daily income was $35,000 in pretax dollars. What should one do with this kind of money? Carnegie formulated a plan:

> Individualism, Private Property, the Law of Accumulation … these are the highest results of human experience, the soil in which society so far has produced the best fruit....
>
> This, then, is held to be the duty of the man of wealth: First, to set an example of modest, unostentatious living, shunning display or extravagance; to provide moderately for the legitimate wants of those dependent upon him; and, after doing so, to consider all surplus revenues which come to him as simply trust funds, which he is called upon to administer, and strictly bound as a matter of duty to administer in the manner, which, in his judgement, is best calculated to produce the most beneficial results for the community—the man of wealth thus becoming the mere agent and trustee for his poorer brethren, bringing to their service his superior wisdom, experience, and ability to administer, doing for them better than they would do for themselves.

Carnegie eschewed the idea of waiting to give money to relatives or charities on one's death as wasteful. What one should really and more nobly do is distribute it properly *during* one's lifetime.

This Carnegie proceeded to do with gusto until every small town in America had at least a library and a church organ. *"Although the busiest bee will receive the most honey,"* he admonished, *"he who dies rich dies disgraced."*

While William Bourn certainly did not have the scope of Carnegie's income disposal problem and probably not quite the same moral fervor, he nevertheless was well known in the community as a man willing to devote his time and resources to projects he felt were important.

One of the amenities of wealth is the ability to join a social club and rub elbows with fellow members one considers of equal class. The venerable Pacific Union Club in San Francisco has long been considered the bastion of proper gentlemen. It was formed in 1889 from a marriage between two other clubs and occupied several different locations in the city.

Following the Great Earthquake and Fire, the club members searched their rolls for an individual who would have the persistence, ability, and vision to build them a grand new club, which had been destroyed. They found the perfect candidate. In April 1908, Bourn was elected president of the Pacific Union Club.

Bourn immediately asked for and was granted special powers to appoint special committees to carry forward the urgent needs of the club. Then he changed many bylaws permitting an increase in indebtedness, disposed of the old location near Union Square, and resolved thorny insurance matters resulting from the earthquake.

The next step was the selection of a new location. The site chosen was the James C. Flood mansion located high atop Nob Hill, which was originally built by the Comstock Silver King in 1886. Flood's neighbors included Hopkins, Stanford, Huntington, and Crocker of Transcontinental Railroad fame. Each built an imposing mansion, and stories evolved about peculiar happenings on the hill.

The Flood property was owned earlier by Joseph G. Eastland, a wealthy San Francisco pioneer. In 1880, an event occurred there that became famous in the annals of the city's eccentric history—the burial of Joshua Norton.

Norton was a small, comic-opera figure who took to wearing military uniforms, printing his own money, and proclaiming himself "Norton I, Emperor of the United States and Protector of Mexico." Roaming the city with his two mongrel dogs, named Bummer and Lazarus, this self-appointed little madman delighted the city with his crazy antics. When he died, city flags were flown at half-mast. The funeral cortege was followed by 30,000 people to the grave that had been dug for him in the Eastland family plot. Fifty-four years later, in 1934, members of the Pacific Union Club discreetly moved the mortal remains of Joshua Norton to Woodlawn cemetery.

The Flood mansion actually was fairly well burned out from the Great Fire, but the outside walls were still standing. Bourn commissioned Willis Polk, by then one of the leading architects of the day, to rebuild the mansion to its present glory.

Polk intentionally retained the nearly ruined brownstone walls of the old Flood mansion, effecting a savings of about $100,000 and keeping much of the building's original appearance. In the half million-dollar renovation job, Polk removed a tower, added a third floor, and constructed two semicircular first-floor wings. The stone used in the restoration was from the same mine in Connecticut that had been used to build the original mansion.

Under the watchful guidance of Bourn, Polk added a beautiful skylighted entrance rotunda, a 38-foot Persian rug, a Pompeiian swimming pool in the basement, a Jacobean oak paneled main dining room, and restored the famous porcelain Chinese punch bowl that had miraculously survived the fire.

On Saturday, February 4, 1911, the club was formally opened. Later, an article appeared in the *American Architect:*

> The modern clubhouse of formal type serves a two fold purpose. It provides a place for retirement and recreation

for its members and guests, and also affords a dignified surrounding wherein may be entertained distinguished visitors that it is sought to honor. Perhaps the more pretentious clubhouse offers to a greater degree than any other structure an opportunity to use ornate features of decorative treatment. A lavishness that would be entirely out of place in the detail or decoration of a home or even many types of public and semi-public buildings to be permissible when employed with care and intelligence in the decorative treatment of the clubhouse.

At the annual meeting on March 31, 1911, Bourn beamed with obvious and well-deserved pride and reminded his fellow members:

> You have built a clubhouse, the character of which creates responsibilities that fall on your shoulders. A clubhouse does not make a club, and the important work of the future is the constant and careful upbringing of the character of the club. It is my belief that the proposed constitution and bylaws will greatly help in that direction.[1]

Today, a modestly sized charcoal portrait of Bourn by John Singer Sargent hangs appropriately on the wall of the business office of the Pacific Union Club. Outside, the sycamore trees planted by Bourn still encircle the building.

The San Francisco Symphony was another important philanthropic interest of William Bourn. After the Great Quake, San Francisco rose from the ashes to first rebuild its homes, then its offices, then the city government, and finally its rich cultural life.

On December 20, 1909, 21 civic minded citizens met in the assembly room of the Mercantile Trust Company on California Street and created the Musical Association of San Francisco. Their aims were "the creation of a permanent symphony orchestra, the building of a great opera near Market and Van Ness, and the establishment of a music conservatory with the University of California."

The 21 men formed the new board of governors of the Music Association. W. B. Bourn, Jr., was listed as President of the Spring Valley Water Company. The primary purpose, of course, was to raise money by soliciting guarantors in the amount of $100 per season for five seasons. After two years of fund raising, the first season opened on December 8, 1911. *"The performance began on time,"* according to the San Francisco *Chronicle, "and before some of the smart set had found their seats."*

In the early years, concerts were given in a variety of halls, such as the Curran Theatre, the Cort Theatre, San Francisco Civic Auditorium, and Stanford University. The Cort Theatre was located on Ellis Street between Stockton and Powell Streets. It seated 1,827 patrons and rented for $150 per concert. The Symphony did not make its home in the War Memorial Opera House until 1932, and later in its present day Louise Davies Hall in 1980.

Bourn took over as the second president between the years of 1912-1916. Several of the issues that Bourn presided over during that time concerned concert-going manners. In December of his first year, the Board of Governors decreed that:

> ... the management has received complaints from the subscribers that enjoyment of the music is impaired by their failure to see performances on account of the wearing of hats by women patrons of the concerts.
>
> For the greater convenience of all concerned it is earnestly hoped that HATS WILL BE REMOVED during concerts, and a compliance with this request will be greatly appreciated by all concerned. In San Francisco local ordinances compel by legal means the removal of head coverings that obstructed the view in places of amusement.

On December 11, 1914, a special notice was given that *"Late arrivals will not be seated during numbers, and those who want to leave before the concert is over are requested to do so before the last number begins."* During the winter of 1916, the Association conducted a survey concerning the lighting of the theater during performances for students who wanted to study the scores. The Dark House prevailed by a vote of 175-45.

The musical format was twofold with both a regular symphony and popular concerts. Symphony concerts provided *"serious music for the leisured and managerial classes on Friday afternoon when they either do not have to work or can arrange not to work. Popular concerts provide entertainment for the working classes on Sunday afternoon."*

During Bourn's tenure, the Association gave 10 symphonies and 10 popular concerts per season. Prices for the combined series for box seats were $300; individual orchestra seats were $2.00 for the Symphonies and $1.00 for the popular concerts. Tickets for the last row of the balcony for the Symphony and popular concerts were 75 cents and 35 cents, respectively.[2]

Bourn also took a great interest in Stanford University. He engaged in fund-raising activities and served as a trustee from 1917 to 1923. He was a great friend of Stanford President Ray Lyman Wilbur, and the two men appeared together on the same platform of many civic occasions. Another major charitable act was the donation of the Muckross property to the Country of Ireland in 1932, but this story is reserved for a later chapter. A very large portion of Bourn's philanthropies will be forever unknown, as Bourn in many cases stipulated that benefactors of his humanitarianism refrain from divulging the sources of their gifts.

After his death, the San Francisco newspapers reported the probation of his Will. Bourn left a reasonable amount of money to his grandchildren, sisters, other relatives, friends, churches, and employees. Hopefully, Andrew Carnegie would have been pleased.

NOTES AND REFERENCES

1. Information on the Pacific Union Club has been largely derived from an informal unpublished history of the Club written in the mid-1970s by a member, Edward H. Clark. The only copy is in the Club library.

2. Information on the Symphony was gathered from the archives of the San Francisco Performing Arts Library, especially its informal history on the years 1911-1986, written by William Huck.

~ 11 ~

Bourn as an Internationalist

Under the presidency of Theodore Roosevelt, the United States began to play an important role in international affairs. In 1908, Roosevelt put on an impressive display of naval power for the world to see when he sent his "Great White Fleet" around the globe. It stopped in San Francisco and aroused the admiration of the local leaders. Bourn's thinking began to take on an international aspect. The result was the Panama Pacific International Exposition of 1915.

Two circumstances converged to produce this highly successful exposition. First, the Great Earthquake and Fire leveled much of San Francisco on April 18, 1906. For three days, the fire consumed 28,000 buildings in four square miles. Some 450 lives were lost, and property damage exceeded $350 million. San Francisco went down for the count, but not for long. The city began rebuilding immediately, and its civic leaders cast about for a way to show the world that San Francisco had risen like a Phoenix from its smoldering ashes. The Exposition was a perfect display vehicle.

Second, in November 1903, Teddy Roosevelt saw a chance to "wave his big stick" and exercise some gunboat diplomacy by intervening in a Columbian-Panamanian revolution. The result was that he "took" Panama and began construction of a canal a year later. For 10 difficult years, the first great government corporation in American history overcame extraordinary health and engineering problems to complete the Panama Canal. It was opened to the commerce of the World on August 15, 1914, on equal terms to all nations. A World's Fair would be a wonderful way to commemorate the opening and, of course, re-establish San Francisco as a major shipping center on the West Coast.

Bourn Named to World's Fair Board

On January 6, 1910, William B. Bourn, Jr., was elected—along with 29 others—to the first Board of Directors of the Panama Pacific International Exposition Company, and immediately became active in its finance committee. On April 18 of the same year, four years to the day after the Earthquake, a great citywide fund-raising event occurred on the floor of the Merchant's Exchange. Bourn, followed by William H. Crocker, made the first financial contribution of $25,000. A frenzy of enthusiasm gripped the crowd and a total of $4 million was pledged that day.

On January 31, 1911, the U. S. Congress pushed aside a challenge from the City of New Orleans and named San Francisco as the official exposition site. Later, on October 13, local newspapers reported the largest banquet ever held to that point in San Francisco—900 dinners at the Palace Hotel. President Taft had come to town to participate in the ground breaking ceremonies. Bourn, seated at the head table, spoke about the "Spirit of San Francisco." He was the third and last speaker before the President.

The next day, Saturday, "Sunny Jim" Rolph, the ever smiling major of San Francisco, with the President of the United States by his side, led a wildly cheering parade, complete with bands and soldiers, to Golden Gate Park where Taft turned over the first spade of dirt. On Sunday, the President attended a large luncheon at the Cliff House, where it is said he endowed the city with its well-known watchword, "San Francisco knows how!"

Ironically, the site was soon changed to a 635-acre piece of unimproved tidelands between Fort Mason and the Golden Gate. Finally, on February 20, 1915, President Woodrow Wilson threw a wireless telegraph switch from the White House and the fair was officially opened.

The 10 main exhibit palaces were built in the form of a huge rectangle into a series of flowered courtyards. Midway in this group stood the spectacular 435-foot Tower of Jewels. It contained 90,000 jewel prisms in various colors that were lit indirectly at night by powerful beacons of light—said by some to "rival the Pharos of Alexandria."[1]

The Exposition was considered to be one of the most impressive and artistically successful world fairs ever held. It was a triumphant procession of grandiose vistas with lavishly planned buildings, surrounded by exquisite gardens. Exhibitions from 29 states and 25 countries were displayed, along with a risque amusement center called "The Zone," and a novelty at the time—an airfield.[2]

Much behind the scenes hard work was necessary to make it all a success. Frank M. Todd, writing in his official history of the fair, says that *"William Bourn was a commanding financial figure in the community, and his personal authority and labors were of immeasurable value in establishing the financial foundation of the Exhibition and in helping to set a high standard of contribution to its stock."*[3]

Today, all that remains is the Civic Auditorium, the restored Palace of Fine Arts, the Yacht Harbor, and, of course, the loose landfill that was to cause major earthquake problems 74 years later. But in its time, nearly 20 million dazzled visitors passed through its gates before it closed on December 4, 1915. It was what George Stirling, one of San Francisco's favorite poets, called "The Evanescent City."

As crowds of party-goers flocked to the Exposition, the dark gray clouds of war were already forming over Europe. In fact, at one time serious discussions took place regarding the possibility of postponing the opening, but plans were too far advanced. On June 28, 1914, a Serbian patriot assassinated the heir to the Austro-Hungarian throne. One by one, the European powers lined up with the Allies or the Central Powers. In the middle of the Exposition, on May 7, 1915, the Cunard liner, *Lusitania,* was torpedoed off the Irish Coast with a loss of 1,200 lives, including 128 Americans. At that point, hard choices had to be made.

Throughout American history, periods of strong democratic reform have often been brought to a close by a major war. This action occurs because such periods infect the populace with a crusading zeal that makes them willing to fight for their ideals. The Progressive Era at the turn of the century inspired many Americans to look at World War I as a holy crusade in behalf of justice and democracy.

President Woodrow Wilson first espoused a policy of strict neutrality. Indeed, his successful Democratic campaign slogan in 1916 proclaimed: "He Kept Us Out of War." General American sentiment favored the Allies, but there were 8 million German-Americans, 4.5 million Irish-Americans who hated the British, and various Poles and Jews who were anti-Russian.

Wilson tried to negotiate a "peace without victory" to keep everyone happy, but events were fast overtaking him. German aggression in Belgium, effective British propaganda against alleged German atrocities, and the renewal of unrestricted submarine warfare by the Germans in January 1917 conspired finally to force Wilson's hand. On April 6, 1917, the American Congress declared war on Germany.

It should not be surprising to find that once Wilson opted for war, his arguments would follow the lofty missionary principles of morality. *"This will be,"* he said, *"a war to make the world safe for democracy ... a war to end all wars."*

Bourn Supports U. S. Entry in WWI

William Bourn definitely was not a pole-sitter on the critical question of neutrality. He was all in favor of America going to war. He began to speak out forcefully on the issue in November 1915 at special ceremonies honoring France and Belgium at the Panama Pacific Exposition (more about this speech in the next chapter).

As the war heated up in Europe, Bourn dreaded the Presidential election of 1916 with Wilson's professed neutrality. In a letter to a friend, Alfred Holman, on January 15, 1916, he asked:

> Is America becoming a country without a man? If America does not give more heed to what kind of country they are making, and forget about how much money they are

> making, the disaster that will overwhelm the United States may be complete. With a President who is 'too proud to fight' with materialists 'too proud to feel,' are we becoming a Nation of Squaws?

Writing to another friend, he continues:

> Can anything be done to bring us back to a true Republican form of government, or, after the war, will the world enter a period similar to the Dark Ages? Certainly not France, and I feel that my eyes will be turned toward that country more than ever. These are gloomy words and ere they reach you, let us hope Hughes will be elected....

After Wilson declared war in April 1917, Bourn changed his tune. In July, he wrote President Wilson a personal letter:

> The hopes and prayers of every true American have been nobly answered by you. Our hopes and our hearts are with you. Your belief that mankind has a birthright to Liberty and Peace as against the belief of a foreign power in Lawless Might leads us to hope and pray that Liberty, Honor, and Justice will be armed with might and made the law of mankind.
>
> If I can be of any service to you or my country, command me. If the organizing of our country should be deemed necessary by you, allow us *(putting in a plug for a fellow Republican)* to commend Herbert C. Hoover as a man eminently endowed to serve such a movement under you.

Bourn, of course, had already been hard at work on the American cause. On December 13, 1915, at the Palace Hotel in San Francisco, Bourn organized and was elected president of an organization called "The Friends of France." At that point, it was becoming clear that Bourn' s real sympathies belonged with France. In fact, he was becoming a full-blown Francophile.

Bourn believed that France was, in a special sense, the sister republic to the United States. The Eagle and the Chanticleer were birds of a feather, as it were. He believed that a love for France enlarged the sentiment of American patriotism going back to Lafayette.

At a dedication of the Library of French thought at U. C. Berkeley on September 16, 1917, the anniversary of Lafayette's birthday, Bourn expanded his concept:

> French thought, France, means everything to this nation. America has much to learn. If the youth of America, if the manhood of America, if the womanhood of America, learn, as in time all will learn these lessons, they have learned all there is in life, for it is from France, above any other nation, that we must learn these three lessons—how to fight, how to live, how to die.[4]

On February 2, 1917, The Friends of France sent off the first American Ambulance unit to the European front made up of Stanford student volunteers who wanted to see the war first-hand. This unit, the first of many, was to be affectionately known on the front as the American Field Service, which saved the lives of many casualties. It should be noted that this activity took place two months before Wilson actually declared war.

On March 29, 1917, at the University Club of San Francisco, another sympathetic organization was formed to be called "The American League of California." Dr. Ray Wilbur of Stanford was elected President and Bourn and Professor Charles M. Gayley of U. C. Berkeley were elected members of the executive branch. The group urged the U.S. to declare war, begin national military service, and send strong letters to many politicians. Bourn, as with the Friends of France, was a major financial backer.

Then, on April 24, 1917, a famous meeting occurred in the Civic Auditorium that was significant to the spiritual history of San Francisco. It was the official "leave taking" ceremony honoring another volunteer ambulance unit, this time with 42 students from Berkeley and 21 from Stanford. Furthermore, four American flags were dedicated and would be taken by the unit to the battle front in Europe. One of the flags was to be given to a similar unit that had already gone "over there" in February. The earlier unit became the first to fly a United States flag on a European battlefield—a flag officially sanctioned by the U.S. War Department.

More than 12,000 attended the ceremony, including 3,500 cadets from Berkeley and Stanford, the mayor of San Francisco, the president of Stanford University, a representative of the governor's office, detachments of the Army and Navy, and the Catholic Bishop of California, who blessed the flags. During the two-hour ceremony, the seats were packed. Many speeches were made, flags were waved, both the French and U.S. national anthems were sung, and Boy Scouts carried Allied flags up the isles.

Enjoying the proceedings immensely, William Bourn delivered a short speech:

> The great battle has been fought. The victory is won. The Soul of America is triumphant.
>
> On the 2nd of April through the immortal words of our President, the nation spoke, and the heart of every true American found peace. You carry to France the flag of our country—for our country—for humanity.
>
> Our flag, the flags of heroic France, of martyred Belgium, of dauntless Britain, cannot be furled until liberty, honor, and justice are made the law of mankind, for to that cause is dedicated everything we are, everything that we have.
>
> We have out-soared the thought of self. Victory is God's.

The "First Flag" was actually presented June 4 on the field in France by Stanford student Arthur Kimber. Unfortunately, Kimber died in action as an aviator on September 26, 1918.[5]

The Friends of France also financed the Lafayette Escadrille, a six-plane squadron, which was sent to France before the U. S. entered the war. In appreciation of his wartime efforts, Bourn received the Silver Medal of La Reconnaissance Francaise in 1918 and the highly prestigious French Order of the National Legion of Honor in 1920.

It has not been recorded what William Bourn actually thought about the outcome of the Great War. It is true that the belated entry into the war by the U. S. saved the Allied cause. It is also true that nearly 8 million of Europe's and United States' most promising youth lay dead in battle.

The reality of war-related violence was brought home forcefully to San Franciscans during the massive Preparedness Day parade in July 1916. A mile-long column was passing up Market Street when a bomb exploded among the sidewalk crowd near the Ferry Building. Six persons were killed outright, three others subsequently died, and more than 50 individuals were injured. Two antiwar protesters were arrested—Thomas Moody and Warren Billings. After a long, bitter, controversial trial, the men were convicted of murder, but they were later pardoned.

At the French pavilion of the Panama Pacific Exposition, Alma Spreckels (the eccentric wife of Rudolph) admired the sculpture of Auguste Rodin. Soon she was off to France to visit the artist and brought back one of the castings of "The Thinker." She presented the statue to the city, and made plans to give San Francisco a museum that would concentrate on French art. It was to be called "The California Palace of the Legion of Honor" and would be designed as a replica of the one in Paris (thus adding fuel to the "Great Museum" wars of San Francisco).

Before the museum was finished, however, it was declared a memorial—not to the Spreckels, but to the California dead who had been killed in World War I.[6] William Bourn must have loved the elegant French addition to his city.

NOTES AND REFERENCES

1. The "Pharos" was a lighthouse on (in ancient times) an island in the entrance to Alexandria. It was one of the original seven wonders of the world.
2. Oscar Lewis, *San Francisco: Mission to Metropolis,* (Berkeley: Howell-North, 1966), page 222.
3. Frank M. Todd, *The Story of the Exposition, Vol. I,* (New York: G. P. Putnam, 1921), page 112. These volumes are the official history of the Exposition.

4. The Bancroft Library at the University of California, Berkeley, contains a large collection of private papers of the Bourn family. A collection of notes on Bourn's involvement in World War I is unsigned, but may have been written in part by Professor Charles M. Gayley, co-founder of the American League of California.
5. The Green Library at the University of Stanford contains two books that are helpful in understanding Bourn's involvement with the Friends of France: (a) *The Story of the First Flag,* published by the Friends of France in 1920; and (b) *The History of the American Field Service in France,* published by Houghton Mifflin Company in 1920.
6. Stephen Birmingham, *California Rich* (New York: Simon and Schuster, 1980), page 85.

~ 12 ~

Bourn as a Philosopher

Several of Bourn's associates described him as a philosopher, not so much in the formal sense, but rather as a man who carefully thought out and deeply felt the principles and ideals in which he believed. He also possessed the fervor and ability to articulate his philosophic beliefs publicly.

The core of his thinking on the subjects of the previous three chapters is clearly illuminated in his well-remembered speech at the Panama Pacific Exposition. The date was November 27, 1915, and the occasion was a special day honoring the countries of France and Belgium. Assembled citizens first paid their respect to the Statute of Lafayette in the colonnade of the Palace of Fine Arts. Then they marched to the Pavilion of the Republic of France. All joined in singing the Star-Spangled Banner and the La Marseillaise. Following an introduction by William H. Crocker, Bourn began:

> Many Americans have read, felt, and understood the story of poor Philip Nolan ... the story of a man without a country. There is not an American who can read the story without feeling the emotion which is the purest and noblest of all emotions, which stirs the souls of men—the love of one's country.
>
> The soul of France was born and dwells in Idealism. It was from the germ of French idealism that sprang the soul of America. Idealism has little in common with Materialism.
>
> The flag of France—the flag of America—the Marseillaise—the Star-Spangled Banner—are outward expressions of a soul that is and always will be free. They are the battle

cries of Liberty and Freedom. They represent the struggle of the ages—symbolic to the soul of mankind.

Mankind today consists of two classes. One class to whom the spirit, the soul, is all. It is Idealism. Another class to whom the brain is all … it is Materialism, a people without a soul.

Philip Nolan learned what it was like to be a man without a country. When Materialism breaks itself against the mighty walls of an ideal, a nation will be saved from being a people without a soul.

Efficiency, wealth, material comfort appeal to all; but they cannot produce the glory that France has now found, the glory that France now dwells in.

France is fighting not only for the defense of home, but she is fighting above all for the cause of humanity, for that human civilization which is the growth of conscious altruism.

France is fighting for honor, for country, and to maintain the righteousness of her civilization.

Her glory lies in greatness of heart … in that inward freedom which has the power to understand, feel with, and, if need be, to help others. It is the true foundation of justice, sacrifice, love. Without it, effort has no special significance.

When America was a youthful giant, without the strength of manhood, she fought one of her battles for an ideal. France understood. France helped. America is now the mightiest among nations. Civilization and humanity lie bleeding. Is America neutral? *(At this point, the crowd roared NO—NO!)*

In what does the might of America consist? Without that freedom of spirit which, unconsciously, gives the right to every American to stand upright before his fellow man, man to man, the might of America is less than nothing. The state guarantees to every citizen the heritage of freedom; the right to feel he is a king; and it is the duty of the citizen to justify his right to kingship.

Every American should feel that the responsibilities of the state rest with the individual. If the state fails in its honor, its virtue, or its duty, that failure affects every individual member of the state and endangers the very foundation.

What America, as a nation, has not done in the last 16 months is now history. As a nation we have remained silent.

Twice in our history our soul has found expression. Let every American search his own heart; let every American search his own soul. Has prosperity weakened us? It is eternally true that through suffering alone men and nations find their greatest selves. As we have sown, so shall we reap.

The mind of the American nation is not satisfied. The heart of the American nation yearns. God grant that they may yet find expression.

Representatives of glorious France, representatives of heroic Belgium … take back to your countries the heart of every true American. America is not neutral. Go where you will … feel the heart of the people of America, and you must know that America cannot be neutral.

A hearty round of applause greeted Bourn as he sat down. It must be remembered that this rather personalized interpretation of the Declaration of Independence occurred almost a year and a half before the United States entered the war.[1]

Another method of examining a man's philosophy is to discover his preferences for reading material. Bourn left just such a record. In 1926, he gave as Christmas presents to friends a collection of his favorite poems, music, and books. This small booklet was privately printed by John Henry Nash, the distinguished San Francisco publisher.

First came the poems, printed in full: *Dickens in Camp* and *The Spelling Bee at Angels Camp,* by Bret Harte; *Jim Bludso of the Prairie Bell* by John Hay; *My Star* and *By the Fireside* by Robert Browning; *John Anderson* by Robert Burns; *Psalms 15, 19, 23, 27; Recessional* by Rudyard Kipling; and *Crossing the Bar,* Alfred Tennyson.

A copy of a letter about the value of friendship was next, written by Clarence King to Horace Cutter in 1888, both good friends of the family.

A third heading is music: *La Boheme* by Puccini; *Tristan Und Isolde* by Wagner; *Sonata-Nocturnes* by Chopin; symphonies and sonatas by Beethoven; all the works by Bach; and *Symphony Pathetique* by Tchaikovsky.

Finally, Bourn listed his favorite books: *the Book of Isaiah; Tale of Two Cities,* treatment of the French Revolution by Charles Dickens; *Lorna Doone,* a Victorian story of courageous people and love under duress by R. D. Blackmore; *The Man Without a Country* by Edward Everett Hale, and used in Bourn's speech in 1915 at the Panama Pacific Exposition; *Adam Bede,* about English pastoral life, by George Eliot; *Kidnapped,* a tale of high adventure, by Robert Louis Stevenson; *Misunderstood* by Florence Montgomery; *Kim, They,* and *The Brushwood Boy* by Rudyard Kipling, storyteller of colonial England; and *Wuthering Heights,* a dark impressionistic tale of revenge by Emile Bronte.[2]

The Spring Valley Water Company was located at 425 Mason Street in the building still used today by the San Francisco Department of Water. At that office, Bourn published an occasional company magazine called *Spring Valley Water,* edited by Edward F. O'Day. It was sometimes used for company propaganda, and now and then Bourn would write a Philosophical article that suited his fancy, such as "San Francisco, the City of Promise," in 1922, and "Thoughts on Character," in 1923. A 1928 edition was completely devoted to a tribute to the San Francisco poet, George Sterling.

NOTES AND REFERENCES

1. This speech is quoted from the Bourn papers already mentioned at the Bancroft Library in Berkeley.
2. A copy of this small book may be found at the Bancroft Library in Berkeley, the Green Library at Stanford, or Filoli.

~ 13 ~

BOURN AS A BUILDER

As a man of substantial means, William Bourn had the luxury of indulging his fantasies with various real estate projects. For millionaires of his era, it was important to make a statement in construction—not only with size, but also with architecture. Bourn was no exception.

The story of Bourn properties really begins with William Bourn, Sr. His first real home was at the comer of Third and Brannan in San Francisco. His second home was at 1105 Taylor Street, a rather large Victorian complete with a "widow's tower" on the roof.

Between 1868-1872, Bourn, Sr., bought a cottage and vineyards for his wife, Sarah, on a site located 2 miles southwest of St. Helena. Soon, they tore the cottage down and replaced it with an elaborate Victorian with two large wooden towers. Unfortunately, it burned to the ground in 1888, several years after the death of W. B. Bourn, Sr.

The stone arch ruins stood until 1895, when Sarah began construction of a new house, built largely of masonry. It was completed in 1903-1904. This house also was called Madroño, which is chronicled in Chapter 4. Chapter 5 describes construction of the nearby Greystone Winery, which W. B. Bourn, Jr., undertook the same year his mother started rebuilding her St. Helena residence.

In 1892, Sarah purchased a large home at 2030 Broadway for herself and her daughters—Ida, Zaidee, and also Maud, who preferred to stay in San Francisco rather than endure the isolated and difficult life at the ranch near San Luis Obispo. Just two years earlier, W. B. Bourn, Jr., decided it was time to build a proper townhouse for his family and commissioned Willis Polk to do the work on the plot he bought at 2550 Webster Street. The result was a very unusual mansion by a very talented architect.

The Bourn mansion revealed bold and distinctive classical details, with a powerful variety of Georgian forms. Sally Woodbridge, an important California architectural historian, described the mansion in detail:

> At the time it was built the house—which virtually fills the lot—was visible on three sides. Polk articulated the back part as a succession of separate, gabled roofed sections. The now visible side elevation has a picturesque medieval aspect that reverses the Classical formality of the front.
>
> Polk's design gave the street facade monumentality both by raising the pianonobile (main story) a floor above the street level and by over scaling the roof cornice, dormer windows, and chimneys. Instead of a central entrance sequence with a grand stair, a window set in a boldly scaled architrave (frame) provides the main focus.
>
> Below the balustrade (railing), an unassuming passage cut through the rusticated basement wall leads to the entrance. In another departure from the academic Georgian Revival style, Polk enlivened the texture of the walls by combining Classical detail in sandstone with clinker brick.[1]

Today, the mansion is surrounded by consulates and high-rise apartments, looking rather forlorn and out of place with its rough surface and precariously placed multiple chimneys.

Willis J. Polk (1867-1924) was not only Bourn's official architect, but he also played a major role in the rebuilding of San Francisco. Polk first worked for Daniel H. Burnham in Chicago. He was then placed in charge of Burnham's San Francisco office after the fire in 1906. In 1910, the office was turned over completely to Polk and renamed Willis Polk and Company. He exerted a major influence on the architecture of the Panama Pacific Exposition.

A good example of Polk's imaginative spirit and willingness to experiment was the Hallidie Building at 130 Sutter

Street. Named after the inventor of the cable cars and completed in 1918, it was the world's first glass-walled structure. It was also a low budget, fast construction building, originally painted blue and gold for its owner, the University of California, and boasted cast iron ornaments, friezes, and elaborate fire escapes.

The faith of the enthusiastic and dramatic Polk was expressed by his friend Bruce Porter in a eulogy at Polk's funeral: *His vision, to the last, was always of this city of San Francisco as the most noble architectural opportunity of the New World.*[2]

One of the most striking examples of the new spirit of San Francisco architecture at the turn of the century was the so-called "Burnham Plan." It was put forth by Daniel Burnham, the leading City Beautiful Movement planner in the U.S. The idea was to leave the eclectic chaos of the old Victorian city behind and create a new image of classical order and grandeur. Its planners were strongly influenced by the Ecole des Beaux-Arts in Paris, the leading architectural school in the world at that time.

The plan called for extensive street lighting, large parks, and grand boulevards. All these streets were to converge, like a spoked wheel, at Van Ness Avenue and Market Street, where a great Civic Center would be erected, rivaling the great city of Paris itself. The plan was presented to the Board of Supervisors the day before the Great Earthquake. The graft trials of Schmitz and Ruef further distracted the city, and the plan never was implemented. San Francisco did, however, manage to get at least a civic center, which many critics have called the most magnificent collection of neoclassical municipal buildings in America.

In 1896, William Bourn, Jr., reacquired control of the Empire Mine in Grass Valley. Well satisfied with the job that Willis Polk had done for him on Webster Street, Bourn then commissioned him to build an English-manor style residence on the mine property. Built in 1897-98, and intended as a summer residence only, this lovely house was called simply "The Cottage" to differentiate it from the other homes the Bourns owned.

Surrounded by 12 acres of formal gardens, fountains, and a reflecting pool, the home contains four large bedrooms and two baths on the second floor. The downstairs exudes a feeling of comfort and charm with rich heart-of-redwood paneling. An extensive collection of rose bushes was personally cultivated by Mrs. Bourn. In fact, the only problem with this idyllic setting was the proximity to the loud noises of the mining operations. It was said that when the Bourn's daughter, Maud, would visit the cottage, she pleaded with her father to stop the stamping operations for a couple of days.

Bourn continued his building program at the Empire. A new office building with facilities for the mine manager was constructed, plus new office facilities for mine supervisors and a map room.

Realizing the important contributions that his trusted manager, George Starr, made to the running of the mine, Bourn built a smaller version of the "Cottage" for him. It was called the "Ophir" cottage. Only the foundation is now visible, because fast-moving flames destroyed it in October 1935.[3]

A clubhouse was also constructed in 1905, with facilities for squash, tennis, bowling, dancing, and several bedrooms. It was designed to serve as a social club for management and a place to entertain visitors. The entire property—land and mining complex—was turned over to the State of California in 1975, and is now operated as a State Historic Park.

As a result of the Great Earthquake of 1906, the vagaries of the San Francisco climate, and the availability of more spacious land, many wealthy people began relocating their homes south of San Francisco on the peninsula. Although the house of Webster Street had escaped damage from the earthquake, the Bourns followed their friends, renting a palatial estate in Burlingame called "Sky Farm" from the Crocker family.

Bourn again turned to Willis Polk, only this time he was commissioned to build a grand country estate 25 miles below San Francisco near present day Woodside. The result was "Filoli." In an age of conspicuous consumption, when a man's home was expected to reflect his station in life, Filoli stands as an impressive monument to money, style, and comfort.

The name Filoli was coined by Bourn from the first two letters of his motto: **FI**ght, **LO**ve, and **LI**ve.

Construction began in 1915, and the Bourns moved into their nearly completed home in September 1917. Polk reportedly drew 19 different designs and asked Bourn to select the one he preferred. The site was chosen because of its proximity to the holdings of the Spring Valley Water Company and its resemblance to the Lakes of Killarney in Ireland.

The estate contains 654 acres, with the historic mansion surrounded by 16 acres of enclosed gardens. It remains intact in its original setting because it lies within the protected watershed of the Crystal Springs water system. The house has 36,000 square feet of interior floor space, 43 rooms (excluding baths, closets, and storage space), 17 fireplaces, and 11 chimneys. The style can be considered a more conventional Georgian Revival than their city townhouse. The doors are French, the exterior brick is laid in Flemish bond, the trim suggests the Stuart period, while the tile roof reflects the Spanish tradition.

The gardens were designed by Bruce Porter and planted by Bella Worn. They are a succession of separate areas, each with a distinct character, such as the Sunken Garden, the Pool Pavilion, and the Yew Allee. The gardens are designed to take advantage of the natural surroundings. The Tea House and the Carriage House form a focal point in a successful blend of the formal and the natural.

The house was estimated to have cost at least $500,000. Before the house was completed, Polk and Bourn came to a severe disagreement over the additional costs to complete Filoli. Bourn ended the friendship, and with it, Polk's involvement in Filoli. He decided that the boisterous architect's extravagant habits would lead to Polk's ruin. He then put the commission into a trust for Polk's wife and appointed Arthur Brown, Jr., to finish the job.

Many articles and books have been written about the grandeur of Filoli. The best way to experience the estate is to take a leisurely trip down Skyline Drive (Interstate 280) south of San Francisco to the mansion itself. It is now owned by the National Trust for Historic Preservation and is open for tours.[4]

One remarkable story remains to be told about Bourn as a builder. Between 1926-29, Bourn commissioned Arthur Brown,

Jr., to produce, in effect, an urban monument to himself. His final design included a seven-story triangular base, topped by a 31-story pentagonal tower with an elaborately sculptured top. The building was similar in concept to the newly completed Chicago Tribune Building, which resulted from a worldwide design contest. The Chicago tower supported a wild assortment of Gothic flying buttresses on its top.

A San Francisco *Chronicle* architectural writer described Bourn's final building scheme this way:

> If the International style of building is a sort of intellectual Teflon, shedding past influences, Brown's Bourn Tower represents a sort of intellectual Velcro, holding fast to whatever in past styles it finds sympathetic. Brown's design was a fine example of the 20s eclecticism much admired by the Post-Modernists.[5]

Bourn's nephew, F. B. Hayne, says his uncle paid $1 million for the property because his father had once owned it. Bourn's grandson, Arthur William Bourn Vincent, suggests that he probably paid too much for it and suggests another motive: *In those days, everyone wanted a big building on Market Street—the Floods, the Hobarts, the Crockers. He wanted to put one up bigger than anyone else.*

The tower was never built for the same, sad personal reasons previously mentioned that prompted Bourn to sell his water, gas, and mining companies.

NOTES AND REFERENCES

1. Sally B. Woodridge, *California Architecture,* (San Francisco: Chronicle Books, 1988) page 68.
2. *Here Today, San Francisco's Architectural Heritage,* sponsored by the Junior League of San Francisco (San Francisco Chronicle Books, 1968) page 330.
3. From the *Morning Union,* Grass Valley and Nevada City, October 30, 1935.

4. One of the best examinations of the mansion and its gardens is *Filoli,* compiled by Timmy Gallagher under the auspices of the National Trust for Historic Preservation, revised 1990, and available at Filoli.
5. Michael Robertson, "In the Beginning," San Francisco *Chronicle,* February 19, 1986.

   Another book helpful in understanding the history of San Francisco's architecture is *Splendid Survivors,* The Foundation for San Francisco's Architectural Heritage, 1979.

~ 14 ~

THE FAMILY AT HOME AND ABROAD

Over the years, the family became increasingly active in the social life of the city. The San Francisco blue book for the 1909-10 season lists family affiliations with various clubs. Bourn belonged to the San Francisco Golf and Country Club, the University Club, Pacific Union Club, Bohemian Club, and Burlingame Country Club. Bourn's daughter, Maud, and his sisters, Maud and Ida, belonged to the Town and Country Club and the San Francisco Golf and Country Club. Mrs. Bourn was a member of the Francisca Club and the Town and Country Club.

The Bourns traveled extensively. On one of their Atlantic crossings, the Bourn's daughter Maud met and fell in love with Arthur Rose Vincent, the second son of Col. Arthur Hare Vincent of Summerhill, Cloolara, in County Clare, Ireland. Vincent had recently been appointed judge for his Majesty's Court for Zanzibar. Bourn, however, was very opposed to the thought of his daughter living in Zanzibar. Arthur Rose Vincent promptly solved this problem by resigning from the judicial service of the British Foreign Office.

Maud and Arthur Vincent were married March 30, 1910, at St. Matthews Church in San Mateo. The wedding reception was held at Sky Farm. Eight months later, in November 1910, Bourn purchased the vast Muckross estate on the lakes of Killamey consisting of 13,000 acres with an elegant Elizabethan-style house. He presented this estate to his daughter and son-in-law as a belated wedding present.

Maud and Arthur Vincent had two children, a daughter, Elizabeth Rose, born January 13, 1915, and a son, Arthur William Bourn Vincent, born July 16, 1919.

In 1915, Bourn purchased 1,800 acres of land south of Crystal Springs Lake adjoining the Spring Valley Water Company property. Setting aside 654 acres for his Filoli estate, he sold the remaining acres to the Spring Valley Water Company. Construction of Filoli (see Chapter 13) was started immediately and was sufficiently complete to permit occupancy in 1917, although gardens, tennis courts, and other amenities were not completed until the 1920s.

Mrs. Bourn, who loved gardening, worked enthusiastically beside Miss Isabella Worn, a noted gardener and floral decorator, in planting the Filoli gardens. In later years, the Bourn's utilized Bella Worn's talents to create floral themes for their parties at Filoli, which were often lavish with floral decoration.

Like her husband, Mrs. Bourn strongly supported the Allied cause long before the United States entered World War I. After returning from abroad in late 1916, where she had spent most of the summer with her daughter and son-in-law at Muckross, she held a tea at the Palace Hotel for her San Francisco friends.[1]

She reported on the work being done at the "California House" in London, which flew the Bear flag to notify disabled Belgian soldiers that they would find a haven there. She told of the 200 disabled soldiers sheltered in California House, where they were learning to speak English and where they were taught the various handicrafts by which men in their condition might make a livelihood when they were ready to leave. Mrs. Bourn noted that the house had provided more than 10,000 meals to soldiers in the past few months before her return home.[2]

The money for California House was raised with very little publicity. For the most part, Californians in England were the principal donors. Among the contributors were Mr. and Mrs. Arthur Rose Vincent.

At the same time that California House was providing these services, Mrs. Herbert Hoover was also busy in London working on relief for war-stricken Belgium and helping to organize and manage the American Women's Hospital for wounded British soldiers.[3]

Some time later, the Hoovers were guests of Mr. and Mrs. Bourn at the Grass Valley Cottage. Hoover was delighted with the opportunity to recall his days in Nevada City where, as a young mining engineer just out of Stanford, he had been employed at the Reward Mine.

In general, when they were not traveling the Bourns entertained frequently and in grand style. Filoli was alive with activity. Bourn held tennis tournaments and spared no expense on his parties. For big parties, a tent covering the front courtyard was erected. After dining in the tent, the guests then enjoyed an evening of dancing in the adjacent ballroom, which was designed specifically to facilitate entertainment of large groups.

Filoli was open most days for afternoon tea for any of Mrs. Bourn's friends who wished to join her.

On a less lavish and smaller scale, the Bourns made use of the Cottage at Grass Valley. These were more intimate gatherings, especially for those friends who were not so easily distracted by the constant clatter of the mine's stamp mills.

Although months might pass without a visit from the Bourns, the Cottage was kept immaculate by a jewel of a housekeeper, Katherine Moriarty (1865-1956). Katy became an important asset to the Bourn family, working as a caretaker and housekeeper at the Empire Cottage for 34 years, until her retirement in 1934. The Bourn's grandson, Arthur William Bourn Vincent, who frequently visited the Cottage as a child, is said to have remarked: *"Although the Bourn's may have owned the Cottage, Katy ran it. There were always cookies available for young children, but no one dared set foot in the kitchen area."*

Katy was deeply religious, and either George Starr or Sam Eastman drove her to church. When they were not available, they arranged for her transportation by others. The Bourns and the mine management appreciated Katy's contribution to life at the Cottage. It was a solitary life, however, and for many long weeks each year, Katy strolled the holly-bordered paths of the Cottage alone with her thoughts.

More than just a charming hostess, Mrs. Bourn was a considerate, thoughtful, and generous person, greatly appreciative of any service or kindness she received. An insight into her caring, sympathetic nature is revealed in the following letters to Mrs. Amy Bull in 1928. Mrs. Bull was a registered nurse who attended Mrs. Bourn for six months during an illness in 1926-27:

Dear Mrs. Bull:

Through Miss Carlson last evening I heard of your sorrow—I know you and this sister were very close and I am sorry that this great grief is yours.

Far away in this new country with strangers about you makes it doubly hard—only it is in sharing our sorrows, even more than our joys, that brings us closer to each other, and your husband will be everything to you now.

You are at work Miss Carlson said and doing for others, as is your profession, is helpful too—for it brings a blessed reward, even when those you care for seem to be ungrateful.

You will be glad to hear this naughty patient is better the past months than for years. The last visit to Santa Barbara with the increased diet has set me on my feet, at last Ifeel like doing something.

Our fruit this season is late and not very plentiful, but Iam sending you a few nectarines with my love—and believe me with deepest sympathy.

Very sincerely yours

Agnes Bourn

Sometime previously, on the occasion of Mrs. Bull's marriage, Mrs. Bourn had sent a sterling silver service for 12 as a wedding gift, and a year later remembered the anniversary with the following note on stationery engraved "Filoli House, San Mateo, California."

Dear Mrs. Bull

Off here in the wilds it is difficult to do many things I wish to do, and today I was unable to even telephone you, though often during the day I have thought of you and wished you and Mr. Bull many happy anniversaries. With greetings to you both.

Very sincerely yours
Agnes Bourn

Thus, did Mrs. Bourn express her warmth, her thoughtfulness, and appreciation for services rendered for a short six months in 1926-27. The author thanks Mrs. Bull's daughter, Mrs. Daphne Towns, a docent at the Empire Mine State Historic Park, for permission to print these letters.

NOTES AND REFERENCES

1. *The Morning Union,* Grass Valley, California, October 17, 1916.
2. At the outset of World War I, Germany struck at France through tiny Belgium. Unprepared, Belgium fought courageously, but suffered monstrous casualties. Many of the Belgian wounded ended up in England because their homeland had been occupied by the Germans and there was no place there for them to be cared for.
3. An unidentified clipping, which appears to be from a Stanford University article concerning its famous alumna, Mrs. Herbert Hoover.

~ 15 ~

The Muckross Story

Ask any travel agent or the man in the street in Dublin where the most beautiful and greenest place in Ireland is and Killarney in County Kerry will surely be high on the list. This picturesque town is surrounded by MacGillycuddy's Reeks, the highest range in the country. High in these mountains, a traveler crosses the Gap of Dunloe. The Gap, a rugged gorge 4 miles long, is best traversed by the local form of transportation—a brightly colored pony cart, where passengers sit back to back.

The long trek leads down to the shores of the Lakes of Killarney, where the tour continues by boat. The scenery around the lakes—thick forests, stark crags, and enchanted islands—could hardly be more romantic. Adventure also exists, with the shooting of the rapids at Old Weir Bridge. The town, with its lovely lakes, is surrounded by the "Ring of Kerry," a circular 112-mile drive past steep hills, quaint villages, and enthralling seascapes similar to Big Sur and the Monterey Peninsula in California.

A magnificent estate stands on the edge of the lakes called Muckross House. The ancient ruins of Muckross Abbey are nearby on a peninsula. This historic home also represents a love story within the Bourn family. In 1910, William Bourn bought and gave the estate as a wedding present to his daughter, Maud, and son-in-law Arthur Vincent, an Irish nobleman.

The property has a long history. By the time the Anglo-Normans reached Kerry around 1200 A D., the MacCarthy's were substantially in command there. In May 1588, the heiress to the Muckross land, Lady Ellen MacCarthy, secretly married her cousin, Florence McCarthy. The story is told that the bride crossed the lake by moonlight for a midnight ceremony at Muckross Abbey. The marriage angered the English Crown, which preferred a different suitor. Florence was arrested and imprisoned for some time in the Tower of London, but was later released and allowed to recover his lands and assume the title of McCarthy Mor.

Muckross passed to the Herbert family in the 1700s. The present house was built in 1843 in the Elizabethan style for Henry Arthur Herbert, a member of parliament for County Kerry. The architect was William Bum, who had designed many houses for the nobility of England.

Interesting features include its mullioned windows and steeped gables topped by pointed pinnacles. Portland limestone was used for the facing, and this was shipped from England to Kenmare and thence by cart over the mountains to Muckross. The house contains 100 rooms, including 25 bedrooms and 62 chimneys. The cost was extravagant in its time, costing more than 1.5 million pounds in current values.

A succession of famous visitors followed—Empress Eugenie of France, British statesman Arthur Balfour, Irish author George Bernard Shaw, and William Butler Yeats, who composed a poem that can still be seen in the old visitors' book. Queen Victoria and her husband, Prince Albert, visited in 1861 with their children. A magnificent Persian carpet and tapestries were especially made for her visit.

The Muckross Estate was bought by Lord Ardilaun, a member of the Guiness family, in 1899. In turn, he sold it to the Bourns in 1910. Maud Vincent's tragic death occurred in 1929. The Great Depression came and it was costing more than $65,000 a year just to run the place with its army of gardeners and servants. So, in 1932, the Bourns and Senator Arthur Vincent presented this lovely house and 11,000 surrounding acres to the Irish nation as a memorial to Maud Bourn Vincent.

Between 1932 and 1964, the house was not lived in. At that time, the furniture was taken to furnish the residence of the President of Ireland and various Irish embassies. Then, in 1964, Muckross was made available to a group of local trustees, who opened it to the public as a museum to record the social history and folk crafts of the people of Kerry County. In recent years, the land has been extended to include 25,000 acres and it is now known as Killarney National Park.

Thus, what started out as a wedding present for Maud and Arthur Rose Vincent is today a beautiful memorial park, dedicated to the preservation of the history of Kerry County—truly an example of philanthropy in its highest form.

William Vincent, Arthur's son, remembers Muckross, where he was born and where he lived in the 1920s as a boy:

> There was no electricity and no heating. Everything was done by fireplaces, peat stoves, candles, and gas lamps. Labor was cheap in Kerry in those days and readily available. There were 85 servants outside the house itself. There were gardeners, farmers, gamekeepers, and the like. In Muckross itself we had four servants in the kitchen, four in the pantry, four maids, and my father's valet.
>
> The architects of the 19th century aristocrats who built the place had ideas which might strike us today as a bit impractical. There is a tremendous distance in terms of stairs and winding corridors, from the kitchen to the dining room. This was in the plans so that the odor of cooking should never reach the nostrils of those at the table.
>
> During the 1920s, the great days at Muckross were the months of December and January, when guests came from all over for the woodcock. There would be six guns and a lot of beaters. There was hunting also for snipe, pheasant, and duck. Deer stalking was the sport of the fall. There were no limits except those set by the owner.[1]

Upon entering the main hall of Muckross, a visitor is introduced to some aspects of the life of its former owners. The walls are decorated with trophies of the chase. Native Red Deer and Japanese Sika Deer are well represented by many heads and antlers, along with two heads of wild goats from nearby Tore Mountain. Over the fireplace are the huge antlers of a Great Irish

Deer, extinct for more than 10,000 years. From point to point, they are 8½ feet wide, and the animal would have been about 9 feet at the shoulders. Various other rooms contain portraits of the Bourns, Vincents, and Lord Ardilaun.

No mention of Muckross House would be complete without mentioning the gardens. Much of the development is due to the personal interest of Arthur Vincent. Among other features, the gardens contain a stream garden, many Rhododendrons, an arboretum, a rock garden, and a bog garden. The design of the gardens is informal, with large expanses of lawns within a woodland setting. The distant vista of lakes, mountains, rivers, and waterfalls is hauntingly magnificent.[2]

As a nostalgic reminder, scenes from the Muckross estate were included in the murals of the ballroom at Filoli. Various cuttings also were brought across the ocean to be grown at the Grass Valley Cottage and Filoli, thereby ensuring a link among all three properties.

NOTES AND REFERENCES

1. Charles McCabe, "Worth a Journey," the San Francisco *Chronicle,* June 23, 1976.
2. Information on Muckross House has been gathered from various guide books sold in the bookstore at the Killarney National Park.

~ 16 ~

The Twilight Years

As noted previously, the Bourns moved into the Filoli mansion in September 1917. Work continued on unfinished projects in the gardens and grounds, as well as on internal details of the house, particularly the ballroom. But the estate was eminently livable by September. The Bourns settled into the comfortable and luxurious surroundings and resumed their busy lives.

Four years later, however, the lifestyle of W. B. Bourn was abruptly altered when he suffered a debilitating stroke in August 1921 at Filoli. He recovered to a considerable degree, but just over a year later, in October 1922, at age 64, he suffered a paralyzing stroke while he was at the Empire Mine that left him confined to a wheelchair for the remainder of his life.

At the time of the second stroke, Bourn went to considerable lengths to keep the state of his health private, as demonstrated by an article in the *Grass Valley Union* on October 7, 1922:

> The San Francisco papers of yesterday morning carried articles stating that the state of health of W. B. Bourn, who is spending some time at the Empire Mine, was alarming. *The Union* was informed of the current rumors in San Francisco, but declined to print them in the absence of confirmation. Yesterday, S. P. Eastman, Manager of the Empire and a close associate of Mr. Bourn, issued the following statement.
>
> *Today's San Francisco newspapers printed a report that Mr. W B. Bourn, President of the Spring Valley Water Company and Empire Mines, was stricken yesterday. The report*

is wholly without any color or foundation. Mr. Bourn is at the Empire Mine in Grass Valley on a vacation and business trip. He is taking a great interest in the operation of the Mine and held a conference yesterday on mine affairs.

Mr. Bourn has greatly improved since his arrival in Grass Valley and this morning met several of his old associates. Previously, he made a trip to the Pennsylvania Mine.

I know you will be glad to inform the public through your paper that Mr. Bourn has so greatly improved since his arrival in Grass Valley.

Signed– S. P. Eastman
Vice President and Manager
Empire Mines and Investment Co.

Considering the stress of the situation, this may have been as good a statement as possible, but it falls short of a convincing denial. It implies that Bourn was in poor health when he arrived in Grass Valley, but then improved. Subsequent events contradicted the statement.

Whatever the events, the crippling disability in no way affected Bourn's active mind and plans for the future. In fact, it was some three or four years later that he initiated his proposal for a large, impressive office building on Market Street, which kept architects busy running back and forth between San Francisco and Filoli with plans for the structure. Unfortunately, the plans never materialized.

On February 12, 1929, the Bourn's only daughter, Maud, died in New York of pneumonia contracted during an Atlantic crossing with her 14-year-old daughter and 10-year-old son. Maud was buried at Filoli in the private graveyard on a knoll overlooking the valley. (The body of the Bourn' s son, who died in infancy in 1882, was reburied at this site.)

Mr. and Mrs. Bourn survived their own ill health and incapacities with reasonable fortitude, but the loss of Maud was a devastating blow. It marked the beginning of the final chapter of Bourn's business life. Within the next 12 months, he sold the

San Francisco Gas Company to the Pacific Gas and Electric Company; the Spring Valley Water Company to the City of San Francisco; and the Empire Mine to the Newmont Mining Company of New York.

Newmont, a company with worldwide interests, had been studying the mines in the Grass Valley area looking for investment opportunities. (The Newmont name was coined to combine New York, where the company's financial headquarters were located, with Montana, where Colonel W. B. Thompson, who founded the company in 1921, had grown up around the mining camps.)

Fred Searls, Jr., then president of Newmont, wrote to his board chairman, Colonel Thompson, on the subject of this purchase, as follows:

> The Empire has been one of the really successful gold mines of the country, having produced probably $50,000,000 from two ore shoots which are becoming lean and dipping into the North Star ground. This mine has belonged to W. B. Bourn of the S. F. Valley Water Company of San Francisco for more than thirty years, and has never heretofore been examined or offered for sale.
>
> However, after my return to New York, Bourn's only daughter died and his wife became ill; he being an invalid himself, determined to sell and wired me, subject to immediate acceptance, his willingness to sell for $250,000 cash. Fortunately, Simkins had just examined the mine and pointed out certain attractive development possibilities, and enough ore in sight to run for two years and liquidate the purchase price with an adequate profit....

Bourn's grandson, Arthur W. B. Vincent, was only 10 years old when the sale of the Empire occurred in 1929, but he noted in later years that his grandfather seemed to be in a big hurry to sell everything. Following Maud's death, her children stayed with their grandparents at Filoli for some months while their father disposed of the estate in Ireland.

As noted in the previous chapter, in 1932 Bourn and his son-in-law, Arthur Rose Vincent, presented Muckross House and the surrounding 11,000 acres of land (2,000 acres of the original estate had been sold off in the 1920s) to the Irish nation as a memorial to Maud Bourn Vincent. Today, the estate is known as Killarney National Park.

In the same year, 1932, Mrs. Bourn, who had been in poor health for some time, became seriously ill and was confined to her bedroom on the second floor of Filoli with her invalid husband. Arrangements were made for construction of a porch with wrought iron railings off the upstairs sitting room at the north end of the house, where they could continue to enjoy the view of the valley and the sunset over the lake.

Although both Bourns were then confined to their upstairs apartments, they continued to entertain for some time. Bourn' s sister, Ida, acted as hostess for their parties and balls, and in suitable weather Bourn moved out onto the porch where he could watch his guests arrive. The last dinner party given by the Bourns at Filoli was the "Drunks Dinner," November 24, 1933, to celebrate the repeal of the Eighteenth Amendment. Bella Worn decorated the house, and the dinner table had a suitable favor for each guest. Bourn's nephew, F. Bourn Hayne, who was somewhat of a poet, wrote a personal verse on each card.

This final soiree of the Bourns is described in detail by Timmy Gallagher (Mrs. Peter Gallagher) in the 1990 edition of her elegant and definitive manual, "Filoli," published for the use of the docents at Filoli.

Three years later, on January 3, 1936, Mrs. Bourn died at Filoli at the age of 75, and six months later, on July 5, 1936, William Bowers Bourn, Jr., died, age 79. The Bourns now lie on the knoll overlooking the valley, beside their daughter, Maud, and their infant son. In 1891, their granddaughter, Elizabeth Rose, was buried in this same plot, which is still owned by the family.

A copy of the Celtic Cross in Muckross Abbey stands on the knoll with an inscription on its base:

FILOLI – TO FIGHT – TO LIVE – TO LIVE

This was an error by the stone mason, and should have read:

FILOLI – TO FIGHT – TO LOVE – TO LIVE

The family has chosen not to correct this error, at least up to the time of this writing.

When Newmont Mining bought the Empire Mine in 1929, the company also acquired control of the North Star Mine. These acquisitions, along with the purchase of more than a dozen smaller mines, formed a new entity called the Empire-Star Mining Company. This large, combined company, owned by Newmont, continued to operate the mines profitably until gold mining was shut down by a Federal Government order during World War II, from October 1942 to July 1945.

After the war, some of the mines, including the Empire and North Star, were reopened. But the economics had changed. The fixed selling price of gold was insufficient to support the cost operation. Following a year of struggle to find ways to operate economically, the mines were again shut down in 1946, this time by the owners. For 10 years after that, efforts were made on a rather sporadic basis to continue to operate. The mines were kept open in the hope that the Government would raise the price of gold from its $35 an ounce set price up to a level that would permit the mines to cover their costs and provide a reasonable profit. That did not happen, however. Overall, during the 10-year period 1946-56, Newmont operated the Empire-Star mines at a net loss.

In 1956, the mines were closed by a labor strike, and they did not reopen. Steps were taken to remove all salable equipment from the underground workings, and in 1959 the machinery was sold off at auction.

From 1959 to 1975, the Empire and other mines of the Empire-Star group lay idle. In 1975, the Empire Mine was sold to the State of California and became the Empire Mine State Historic Park, open to the public. Tours of the house, or "Cottage," and the grounds are given by trained volunteer docents.

In 1938, Filoli was sold to Mr. and Mrs. William P. Roth, principal owners of the Matson shipping line founded by Mrs. Roth's father. Under the horticultural expertise of Mrs. Roth, the Filoli gardens achieved worldwide renown. She retained the Bourn's original plan for the garden and added several acres of her own design.

In December 1975, Mrs. Roth gave Filoli, its elegant gardens, and 39 acres of land to the National Trust for Historical Preservation. An additional 86-acre parcel was later given to the National Trust. Mrs. Roth also established an endowment of $2.4 million to assist the Trust in the maintenance of the properties. In 1982, Mrs. Roth and her family gave the remaining 529 acres of the original Filoli property to Filoli Center. Open to the public, trained docents conduct tours through the beautiful house and gardens.

Mrs. Roth died September 5, 1985. All of the original Bourn furnishings, which she used for her Hillsborough home, were willed back to Filoli.

Thus, more than 360 years after the arrival of Jared Bourn in New England, the saga of this branch of the Bourn family came to a close.

In a report for the California Department of Parks and Recreation, Sheila Skjeie, a Park Historian, wrote:

> Their traditions were those of wealthy, white Protestants with British backgrounds. The Republican party, the Episcopal Church, private schools, presentation at Court, large homes, all were part of the lifestyle and traditions of the Bourn family.

The Bourn family was a strong force for the good of the community throughout California's Golden Age. They left enduring marks on San Francisco through their contributions to the economic, social, architectural, cultural, and religious life of the community. That same force for good continues on into the present day.